Building Safer Healthcare Systems

Peter Spurgeon · Mark-Alexander Sujan ·
Stephen Cross · Hugh Flanagan

Building Safer Healthcare Systems

A Proactive, Risk Based Approach to Improving Patient Safety

Springer

Peter Spurgeon
Medical School
University of Warwick
Coventry, UK

Mark-Alexander Sujan
Medical School
University of Warwick
Coventry, UK

Stephen Cross
Academy of Medical Royal College
London, UK

Hugh Flanagan
ORCNI
Gnosall, Staffordshire, UK

ISBN 978-3-030-18243-4 ISBN 978-3-030-18244-1 (eBook)
https://doi.org/10.1007/978-3-030-18244-1

This Springer imprint is published by the registered company Springer Nature Switzerland AG
The registered company address is: Gewerbestrasse 11, 6330 Cham, Switzerland

Foreword

Health systems across the world face a continuing challenge to improve safety for the patients in their care. As systems and treatment regimes become ever more complex and patients present with more co-morbidities, this challenge increases. Many improvements have been made and continue through the efforts of dedicated professionals. Often, these are quite specific and the result of a focus on a particular condition or issue. More worryingly and as Chap. 1 of this text suggests, the overall levels of experienced adverse incidents have not decreased to the levels that might have been anticipated. In reality, the rate of incidents has remained at a fairly constant level for many years.

This picture accords with my own previous experience of clinical practice. Retrospective investigations, many with worthy intent, would seek to highlight the cause or trail of events. But over the years, it seemed a similar, marginally different pattern of events would be described. It felt as if we were closing a loop around each incident but not truly eliminating them.

My own instincts suggested there had to be an alternative approach that might improve patient safety in a sustainable way.

Participation in the Safer Clinical Systems project based at the University of Warwick and funded by the Health Foundation confirmed that other models existed but were not prevalent in the health sector. The story of the conduct of this project, the results achieved and the lessons learned are described in this text, thus making them available to so many more people. The critical components of a prospective approach, a focus on risk not harm, a system perspective and underpinned by a holistic interpretation of human factors are all clearly articulated in this text. These principles are common in other safety-critical industries, and while some will claim that health is different, there is a need to adapt the underlying model to the health sector (Chap. 2 provides some guidance on how best to learn from other industries).

The authors of this book go further than simply identifying these principles. They have integrated them into a coherent model—indeed an as an educational syllabus that could provide the model for future training of all health staff (see Chap. 10 for outline of syllabus). This development is one of the most important advances in patient safety for decades and if fully implemented could transform levels of safety for patients in the future. It is a really important text and should be part of the education of all who work in healthcare.

London, UK

Prof. Matthew Cooke
Ph.D., FRCEM, FRCP, FRCS (Ed)
MFMLM Dip IMC Director Clinical
Systems Improvement

Contents

Part I
The Conceptual Underpinning to a Paradigm Shift to Improving Patient Safety and the Emergence of the Safer Clinical System Approach

Chapter 1
Patient Safety: Why We Must Adopt a Different Approach

1.1 Introduction

There needs to be a significant change in the way we think about and approach the problem of patient safety in healthcare organisations. This change may be described as radical or perhaps even a paradigm shift, but it is fundamental and, in our view, necessary as to continue with the current failing model would be perverse. Perhaps the most significant single goal would be to change the way the sector thinks about patient safety, to change the prevailing mindset.

There are of course activities and techniques that follow from adopting a new model. These are described and demonstrated in the applications outlined in Section II. However, as with almost all techniques they have procedures, and can be learned, but without the different mental model, they might not be used, or the contribution they can make to improving patient safety, become apparent.

The recently appointed Health Secretary (Matt Hancock) has set out his vision for patient safety in the NHS. Excitingly and challengingly he talked of making the NHS the safest healthcare system in the world. There were clearly attempts to build upon current approaches with mention of learning from mistakes, incident reporting, getting to the root cause and improving our learning processes. All are important and useful but do not go beyond where we are now. However, encouragingly he also, in conjunction with the new Director of Patient Safety (Aiden Fowler), highlighted the prospect of acting on patient safety risks (Secretary of State 2018). It is important to make progress in this area and as we will describe in the rest of this book how a focus upon this can make a real contribution to improvement in patient safety.

The approach advocated here is not new to most other safety-critical industries, but it is largely so in the health sector. Adaptation to the healthcare environment will need to be appropriate and will necessarily take time to build the critical areas of changed thinking. The argument will be based on three strands:

(a) The unacceptable picture of the current position with regard to patient safety.
(b) The contribution that different disciplines can make to creating a new mindset.

P. Spurgeon et al., *Building Safer Healthcare Systems*,
https://doi.org/10.1007/978-3-030-18244-1_1

(c) The practical application and demonstration of this approach in Section II of this text.

In this introductory chapter, we will focus primarily on the existing patient safety profile across most healthcare systems, and hence the implicit argument for change. Chapters 2, 3 and 4 in Section I provide the conceptual underpinning to the Safer Clinical System (SCS) approach, and the conduct of the initial SCS programme is described in Chap. 5. In Section II, Chaps. 6 and 7 illustrate the practical application of the SCS model, while Chap. 8 represents a specific application to a particular issue (Handover in Emergency Care). Finally, Chap. 10 suggests where we go from the arguments presented and crucially includes the first published version of the new Patient Safety Curriculum representing the culmination of this work and the key future direction for patient safety in the NHS.

1.2 Patient Safety—Where We Are Now

Many prominent researchers and authors in the field of patient safety have described the consistent worldwide picture of deficiencies in the profile of patient safety. Hollnagel et al. (2013a, b) summarise the viewpoint very succinctly saying that '*care is simply not as safe as it ought to be*'. Vincent et al. (2008) reported that 10% of inpatients in the UK were likely to experience some form of adverse event. Similarly, Landrigan et al. (2010) suggested that in the USA, this might apply to every third or fourth admission. These rates are endorsed by Braithwaite and Donaldson (2016) and most importantly reported to apply to a large number of healthcare settings (World Health Organisation 2017). The sense then that the situation is similar everywhere was given great credence by the extremely influential paper by Kohn et al. (2000) who in their report '*To Err is Human*' suggested that the majority of adverse incidents were a consequence of system rather than individual failure. Hollnagel et al's (op cit) suggestions as to how this situation may have arisen reinforce the notion of systemic fault citing a set of contextual pressures such as increasing demand for care from an ageing population, rapid advances in technology producing greater complexity and governmental concerns about rising costs creating pressure on the adequacy of staff resources. Mannion and Braithwaite (2017) describe this type of conceptualisation of the problem as 'safe system methodology'. Spurgeon et al. (2017) concur with the general direction of the argument but point to the evolutionary nature of medical care as unwittingly contributing to the problem. They suggest that much of the healthcare delivery system has grown in a piecemeal fashion, a common practical style where new technologies, drugs, procedures and protocols are incorporated and thereby produce increasingly complex and overlapping layers of delivery systems. Although each specific addition may make a contribution to better care, it also increase the multiple interactions obscuring at times the fundamental design of the underlying system to be as safe as it can be. This cumulative acceptance and inclusion of ways of working can make it feel as if healthcare systems were not designed but came into existence, and moreover present a great challenge to the processes of system redesign.

In a review of approaches to improving safety, Illingworth (2015) makes the telling point that the reasons cited by infamous NHS inquiries ranging over the decades from Ely Hospital in 1969 to Mid-Staffordshire and Morecambe Bay some 46 years later are depressingly similar (Kirkup 2015). The recurring factors include clinical professionals isolated and unaware of advances in techniques, various manifestations of poor leadership, organisational systems not working properly such that simple but crucial checks were absent, many forms of inadequate communication and staff or patient concerns being ignored or dismissed. The underlying reasons for the failure to radically improve patient safety are not really different from the challenges faced by any significant change initiative—namely, the failure to truly embed specific changes such that they gradually decay and practice reverts to that in use previously; the systems-wide nature of some problems that are therefore beyond the scope of particular units; and a lack of awareness of the context surrounding certain initiatives which often precludes their transfer from one site to another. It would be wrong to imply that no progress has been made in improving patient safety. There are well-documented initiatives that undoubtedly have had an impact. For example, there have been reductions in catheter-related infections (Provonost et al. 2006), reduced errors in theatres from the use of checklists (Haynes et al. 2009) and better responses to deteriorating patients (Hillman et al. 2017). Nonetheless, the incident rate persists and as we have seen is stubbornly around the same level it has been for many decades.

The cost to patients, families and carers of unsafe care is almost in principle incalculable. In terms of the system itself, a report to the Department of Health in 2014 (Evidence. nhs.uk) suggested it may be more than £1 billion but probably even greater. This suggests that the system contains so many latent sources of hazards to patients that they continue to manifest in different forms. Dealing with some does not prevent others occurring. As Vincent et al. (2013) conclude the assessment of safety by what has happened (number of reported incidents) does not really tell us how safe care will be in the future.

1.3 Advocated Approaches to a Different Model of Patient Safety

It is fair to say that there are some recent conceptual developments in attempts to respond to the prevailing picture, and indeed one might suggest that before time: the general population, and thereby potential patients, may well have wondered why it has taken so long for new approaches to be proposed. We seem to be functioning within a version of insanity often attributed to Albert Einstein where if we keep doing what we are currently doing we will continue to get what we currently get, i.e. a sustained level of patient harm undiminished by the approaches adopted. Even without detailed specification of alternative approaches, there would seem to be an overwhelming argument for doing something different.

The framework developed by Charles Vincent and colleagues (Vincent, Carthey and Burett op cit) is an important contribution to the debate. The model is centred around the concept of safety measurement and monitoring, and this is supported by a virtuous link of five other components. These are:

(a) Past harm, directed towards what is known and the use of previous incident reports
(b) Reliability—concerned with the degree of system reliability
(c) Integration and learning—focusing on how well an organisation is learning from the past and putting in place initiatives that create improvement
(d) Sensitivity to operations—how far are we concerned that care being provided today is safe
(e) Anticipation and Preparedness—do we know whether care will be safe in the future.

Early attempts to try out the framework with three healthcare providers, coordinated by the Health Foundation, illustrated that nearly all the data collected were concerned with past events. Such data Vincent and his colleagues call 'lagging indicators'. From the perspective of this text and the approach represented by Safer Clinical Systems, this accords with where we would characterise the current focus within the world of patient safety. However, we see the most important contribution of the framework—as well as offering a good structure for organisations to think about how to help tackle patient safety—is in the relatively novel addition of the Anticipation and Preparedness domain. This is concerned with aspects that might make care delivery more or less risky in the future. These aspects are known as 'leading indicators'. This is an important step forward in the way we think about patient safety. However, the model suggests that the leading indicators are sparse and do not really provide very specific or concrete guidance on how providers might go about populating this domain. We would argue that the Safer Clinical Systems approach described here offers a way of not only developing this concept, but of highlighting its crucial role in improving patient safety.

The second important conceptual development in the area is that of the notion of progressing from Safety I to a more sophisticated idea embodied in Safety II. This is neatly articulated by Mannion and Braithwaite (2017) building upon an early description by Hollnagel et al. (2013a, b). The latter suggests that an underlying problem is that safety is defined by what it is not, and by this the authors mean that when an adverse incident occurs it is because the safe care one presumed was present was in fact absent. They argue that important implications follow from this conceptualisation. Firstly, the approach becomes one labelled 'Find and fix' which involves a search to find a particular cause of the problem and to initiate changes that are aimed at driving out whatever went wrong. In a strange way, then we appear to be attempting to learn about safety using a restricted set of infrequent events—as by definition most care provision is successful. This is a necessarily skewed basis for learning, and it is fundamentally a reactive approach based on what has happened before. Hollnagel, Braithwaite and Wears (op cit) have labelled this prevailing approach as Safety I. The key change they wish to advocate is to move from this retrospective focus on failure to base our learning on the much larger pool of events (procedures that go well what they call Safety II). This view recognises the complexity and unpredictability of healthcare settings, and that safe care is effectively provided by the adjustments and flexibility of the care professionals. This description seems to characterise safety as a constant battle by healthcare professionals to use their skills and abilities to overcome the demands and challenges of the environment in which they work. In some systems, such as the current NHS, this may well be an apt description. However, Safety II goes beyond this as to leave it as a constant struggle would place undue pressure on the individual clinician. In fact, Safety II argues for a proactive position whereby care is provided within a system where one attempts to consider what might happen, to work to accommodate the fluctuations and adjust before an incident occurs. The authors suggest that this would provide resilient healthcare and that supporting individuals to adjust and adapt to the development requirements is a key role of organisations.

Mannion and Braithwaite (op sit) take a similar view. They suggest that over time the focus of patient safety has moved from personal error and blame directed to the individual clinician to the idea that the system is the problem. While this has broadened the notion of cause, it has encouraged the use of techniques such as root cause analysis in an attempt to specify and eliminate the particular cause. The approach remains essentially retrospective and as the authors suggest gains in patient safety have been limited. They enhance the notion of Safety I versus Safety II by suggesting that the former considers work-as-imagined or in ideal-type form. In contrast, Safety II operates with the notion of work as it is, where clinicians are constantly required to be flexible in how they deliver care. Learning to do this more effectively is their view of how patient safety may be improved. It is less clear just how this may be achieved although they point to the potential contribution of complexity science (Hawe 2015).

These ideas represent an important step forward as they begin to drive towards a proactive and risk-orientated approach. Mechanisms for achieving this are not really articulated in the Safety II model but we argue throughout the rest of this text that Safer Clinical Systems (SCS) encompasses the conceptual advances but also offers practical tools to achieving the goals. Other authors are increasingly advocating a similar shift in thinking. Macrae (2014) argues that a major threat to patient safety is the lack of attention given to early warning signs of potential problems. He is not in this instance referring to the deteriorating patient but rather what he describes as organisational and cultural indicators. He suggests that the disasters that occur in health systems stem from an inappropriate failure (page 441) where information, communications and warnings are constantly misinterpreted or ignored. He again points to an approach that is oriented to capturing future risks as that most likely to improve patient safety. Bruno and Bracco (2016) built upon this notion by trying to design a tool that would help staff in systematically identifying critical components in their current work activities that may influence failure safety. They suggest that this more participatory style can overcome some of the deficiencies in the reporting culture in health organisations.

The arguments we have reviewed here in this introductory chapter are brought together in the paper (Spurgeon et al. 2017) giving an account of the development of the SCS approach and its implementation. Almost all current approaches are retrospective and therefore in building a safer system the first thing that has to happen to trigger the system is that the patient is harmed or suffers an adverse event. Healthcare delivery systems proceed in the belief (hope) that they are safe—then an event occurs which demonstrates that the system actually was not safe. It is this type of thinking that perpetuates the rate of adverse incidents observed. In considering other safety-critical industries that have improved, it is apparent that two aspects of their approach must be transferred to healthcare:

(a) To be proactive and not wait for the hazard or potential problem to manifest itself
(b) To seek out the risk that exists within the delivery system and eliminate or control it before it results in harm.

In these industries, the focus is upon scrutinising the sources of risk, taking action and thereby knowing the level (acceptable level) of risk that can be tolerated. Healthcare as a sociotechnical system with heavy reliance on human resources adds further aspects of uncertainty. However, we believe that the SCS programme has demonstrated that proactive risk-based approaches can be applied to healthcare and that indeed if radical improvement in patient safety is to be achieved they must be used. Before describing the application of SCS, we provide in this first Section I a brief overview of the theories and models behind the SCS approach—in particular the transferable learning from other safety-critical industries (Chap. 2), the focus upon risk, the contribution of human factors and systems thinking as the well (Chap. 3) as the role of culture and context within patient safety systems (Chap. 4). Our intention here is not to make the reader experts in each of these spheres but to provide sufficient background information that clinicians and other healthcare staff can appreciate the basis of an alternative model of patient safety—to change their mindsets so they can replicate the type of applications described in Section II and thereby help make patients safer.

References

Braithwaite, J., & Donaldson, L. (2016). Patient safety & quality. In E. Ferlie, K. Montgomery, & A. R. Pederson (Eds.), *The Oxford handbook of health care management*. Oxford, U.K.: Oxford University Press.

Bruno, A., & Bracco, F. (2016). Promoting safety through well-being; An experience in healthcare. *Frontiers in Psychology, 7*(1208), 1–6.

Evidence. Nhs.uk (2014). Exploring the costs of unsafe care in the NHS: A report prepared for the department of health. https://www.evidence.nhs.uk/document?

Hawe, P. (2015). Lessons from complex interventions to improve health. *Annual Review Public Health., 36,* 307–323.

Haynes, A. B., Weiser, T. G., Berry, W. B., Lipsitz, S. R., Breizat, A. H. S., Dellinger, E. P., et al. (2009). A surgical safety checklist to reduce morbidity & mortality in a global population. *New England Journal Medicine, 360*(5), 491–499. https://doi.org/10.1056/NEJMsa0810119.

Hillman, K., Nosrat, H., & Braithwaite, J. (2017). RRS & the cuture of safety. In M. A. Devita, K. Hillman, & B. Bellomo, (Eds.), *Textbook of rapid response in systems* (pp. 53–57). Cham, Switzerland: Springer International Publishing.

Hollnagel, E., Braithwaite, J., & Wears, R. L. (Eds.). (2013a). *Resilient health care*. Surrey, U.K.: Ashgate Publishing Ltd.

Hollnagel, E., Braithwaite, J., & Wears, R. L. (Eds.). (2013b). *Resilient health care*. Farnham, U.K.: Ashgate Publishing Ltd.

Illingworth, J. (2015). *Continuous improvement of patient safety: The case for change in the NHS*. London: The Health Foundation.

Kirkup, B. (2015). *The report of the Morecambe Bay investigation*. London: The Stationery office.

Kohn, L. T., Corrigan, J. M., & Donaldson, M. S. (Eds.). (2000). *To err is human building a safer health system*. Washington, D.C.: National Academy Press.

Landrigan, C. P., Parry, G. J., Bones, C. B., Hackbarth, A. D., Goldmann, D. A., Sharek, P. J., et al. (2010). Temporal trends in rates of patient harm resulting from medical care. *New England Journal of Medicine, 363*(2), 124–134. https://doi.org/10.1056/NEJMsa1004404.

Macrae, C. (2014). Earley warnings, weak signals & learning from healthcare disasters. *BMJ Quality & Safety, 23,* 440–445.

Mannion, R., & Braithwaite, J. (2017). False dawns & new horizons in patient safety research & practice. *International Journal of Health Policy and Managemant, 6*(12), 1–5.

Provonost, P., Needham, D., Brienholtz, S., Sinopoli, D., Chu, H., Cosgrove, S., et al. (2006). An intervention to decrease catheter-related bloodstream infections in the I.C.U. *New England Journal of Medicine, 355*(26), 2725–2732. https://doi.org/10.1056/nejmoa061115.

Secretary of State. (2018). Patient safety; no room for complacency. https://www.gov.uk. goverement/speeches/patient-safety-no-room-for-complacency.

Spurgeon, P., Flanagan, H., Cooke, M., Sujan, M., Cross, S., & Jarvis, R. (2017). Creating safer health systems: Lessons from other sectors and an account of an application in the Safer Clinical Systems Programme. *Health Services Management Research*, 1–9. (OCO).

Vincent, C., Aylin, P., Franklin, B. D., Holmes, A., Iskander, S., Jacklin, A., et al. (2008, November 13). Is health care getting safer? *British Medical Journal, 337*(7680), 1205–1207. https://doi.org/ 10.1136/bmj.a2426.

Vincent, C., Carthey, J., & Burnett, S. (2013). *The measurement & monitoring of safety*. London: Health Foundation.

World Health Organisation. (2017). *Patient safety: Making health care safer*. Geneva: Switzerland.

Chapter 2
Learning from Safety Management Practices in Safety-Critical Industries

Making Organisations Safer Through Proactive Risk Management, Safety Cases and Organisational Learning

2.1 Introduction

This chapter describes some of the key lessons from the management of safety in safety-critical industries, which might be applied in healthcare in order to improve the safety of care delivered to patients. Certain safety-critical industries, such as civil aviation and the nuclear industry, suffer very few accidents. Such domains are sometimes referred to as ultra-safe systems (Amalberti et al. 2005). What do these industries do that enables them to remain near-accident-free for significant periods of time?

Arguably, many factors contribute to the success of ultra-safe systems. However, looking across safety-critical industries, it is possible to identify a number of core safety management practices that are accepted and expected: the proactive identification and management of risk, the demonstration and critique of an organisation's safety position (i.e. why do we believe the organisation is safe?), and the commitment to continuous organisational learning.

The aim of this chapter is to provide a brief overview of these safety management practices and to describe lessons for the management of patient safety in healthcare organisations. The transfer of lessons from safety-critical industries to healthcare can often be challenging in practice (Clay-Williams and Colligan 2015; Kapur et al. 2016; Sujan et al. 2017). When transferring and applying lessons from industry to healthcare, it is important to understand the underlying theory, the benefits and the limitations of tools and methods within their original industrial context (Sutcliffe et al. 2017).

The next section gives a brief overview of how safety-critical industries proactively seek out and manage risk. Then, the concept of safety cases is described. This concept is useful to make an organisation's risk position explicit. Subsequently, the importance of establishing strategies to promote organisational learning is discussed. The chapter then looks at some of the similarities and significant differences between safety-critical industries (such as aviation) and healthcare, which need to be under-

P. Spurgeon et al., *Building Safer Healthcare Systems*,
https://doi.org/10.1007/978-3-030-18244-1_2

stood when transferring lessons from one context to another. A summary of the key lessons for healthcare concludes the chapter.

2.2 Proactive Risk Management

A defining characteristic of successful safety-critical industries is that organisations seek out and manage safety risks proactively (Sujan et al. 2017). An extreme example of this practice is the public inquiry into the construction of a new power plant in the UK in the 1980s, which considered the risks associated with the new plant for several years before the reactor was actually built (see Box 1). In everyday practice, proactive risk management is not as thorough as in this case, and it is proportionate to the expected levels of risk. However, the principles of proactively seeking out risks and of explicit and transparent scrutiny are fundamental pillars of successful safety management across different industries.

It is useful to distinguish and discuss separately the concept of risk, methods to assess risk, and the organisational strategies and processes for managing risk, as these are separate albeit interrelated issues.

Box 1. The Sizewell B Public Inquiry

The Sizewell B Pressurised Water Reactor in Suffolk in the East of England was built between 1987 and 1995. Before the decision to go ahead with the construction was given, the operators had to submit a detailed preconstruction safety case to the Nuclear Installations Inspectorate, where they set out precisely what kinds of risks they anticipated, and how they proposed to deal with these. The preconstruction safety case was subjected to rigorous review in a public inquiry chaired by Sir Frank Layfield. The public inquiry sat for 340 days between 1983 and 1985. The final, 3000-page report was published in 1987 and concluded that, subject to a satisfactory (construction) safety case, the construction of the new nuclear reactor was in the national interest.

2.3 The Risk Concept

Concepts of risk aim to model or to represent that an activity could in the future lead to some kind of consequences or outcomes that are not desired. The concept of risk has been discussed from different perspectives in the literature and to date there is no agreed definition of risk (Aven 2012). Aven (2012) and Althaus (2005) give interesting overviews of the historical development of different conceptions of risk.

In healthcare, it is common to talk about the risk of developing a specific type of disease or condition, e.g. diabetes risk, and to identify related risk factors that increase the risk. Risk in this interpretation represents a probability. From an engineering perspective, risk is often regarded as the combination of the probability of an event developing and the severity of the resulting consequences. Including consideration of the consequences is important because an event (e.g. failure of brake on a car or a train) can have outcomes of different severities (e.g. negligible injury to fatality). The ISO 31000 standard on risk management defines risk as 'effect of uncertainty on objectives' (ISO 2009). This somewhat cryptic definition incorporates the notion of uncertainty, which in effect separates the risk concept from the measurement of risk. The earlier engineering perspective proposed probability and severity as both definition and measurement of risk. The ISO 31000 and other more recent definitions define risk through uncertainty related to activities and consequences, but leave the measurement open (Aven 2011). This is important because analysts typically make a number of assumptions or rely on background knowledge when assessing risk. These assumptions and background knowledge can be strong or not so strong, i.e. they have associated uncertainties. These uncertainties can have a significant impact on the assessment of risk but are not usually captured in the engineering perspective based on probabilities (Aven 2011). For example, consider a hazard involving the failure of an automatic train protection system (automatic braking system). Engineers might estimate the failure probability p and the severity of the consequences c. However, the train protection system might be a radically new design, and there might be only limited testing evidence available, which was collected under idealised conditions in a laboratory. Therefore, the existing knowledge to support the value of p would be weak. The failure probability p does not provide any indication of the uncertainty that is associated with its value. A possible way to deal with this problem is to articulate the assumptions and their relative strengths separately from the risk assessment, for example, in the form of a safety case (Sujan et al. 2016) (see the section on safety cases below).

Such technical and engineering perspectives of risk can usefully lead to the development of methods for the assessment of risk, as described in the section on risk assessment below. However, there are also other perspectives on risk, which focus on the dynamic and social dimensions of risk (Althaus 2005; Brenkert-Smith et al. 2013; Kasperson et al. 1988). Such conceptualisations of risk emphasise that risk perception and decision-making about risk are inherently social processes that go beyond technical descriptions of risk. For example, the starting point for Social Amplification of Risk Theory (Kasperson et al. 1988) is the observation that there is frequently a discrepancy between the technical assessment of risk and the public perception and response to risk. Risks that have been assessed as small from a technical perspective, sometimes, can be perceived as much greater and trigger consequences at a societal level that were not considered during the technical analysis. According to the theory of Social Amplification of Risk, this is because the public perception of risk is shaped by a range of information sources and social interactions, of which the technical assessment is but one. A recent, tragic example is the GermanWings flight 9525 accident in 2015, in which 150 people lost their lives (see Box 2). This prompted significant public debate about psychological assessment and monitoring of pilots, and it prompted the European Aviation Safety Agency (EASA) to make recommendations for adopting a two-people cockpit rule. This was not backed by technical evidence or technical risk analysis and shows the discrepancy that can exist between the technical perspective of risk and the wider societal impact.

Box 2: GermanWings Flight 9525 Accident

On 24 March 2015, GermanWings Flight 9525 crashed en route from Barcelona, Spain to Düsseldorf in Germany. The flight carried 144 passengers and six crew members, all of whom died in the accident. After crossing the French border, the aircraft started to descend rapidly from its cruising altitude of 36,000 ft (11,600 m) and contact with air traffic control was lost. After a descent of 10 min, the aircraft crashed into a mountain in the French Alps. Subsequent investigations led by the French Bureau of Enquiry and Analysis for Civil Aviation Safety concluded that the accident was caused by deliberate actions of the co-pilot Andreas Lubitz. The analysis of the voice recorder showed that the captain left the cockpit just after 9:30, about half an hour into the flight. Shortly afterwards, the selected altitude was changed from 38,000 to 100 ft. From 9:34 onwards, once the aircraft had started to descend rapidly and noticeably, the voice recorder evidence suggests that the captain was trying to return to the cockpit by pressing the buzzer initially, and then by banging on the cockpit door. An automatic low-altitude warning was issued at 9:40, and the aircraft crashed at 9:41.

2.4 Risk Assessment

The above discussion illustrates that the concept of risk and the process of assessing risk are two different things. How one assesses risk is influenced by the risk concept adopted, but which concept is the most appropriate will depend on the specific situation.

In UK safety-critical industries, the most common approach for risk assessment is to describe risk qualitatively and quantitatively (as required) and to document relevant uncertainties in a safety case. Risk assessment usually entails hazard analysis to identify and to describe scenarios of interest, and risk analysis to describe and to evaluate the risks associated with the identified hazards. The term hazard is not defined unambiguously but can be regarded as a situation of interest with respect to risk. For example, in the aviation context, a runway excursion (aircraft leaving the runway during landing) is a hazard. The risk analysis of such a situation would try and answer questions about what could cause a runway excursion and describe the likelihood (qualitatively or quantitatively) of such an event happening. It would also aim to describe the consequences of a runway excursion and try to estimate their severity. Together, these analyses describe the risk associated with the specific hazard.

These steps can be supported by a large number of specific methods, such as Failure Mode and Effects Analysis (FMEA) and the extension Failure Mode, Effects and Criticality Analysis (FMECA) (IEC 2006), Hazards and Operability Studies (Hazop) (Kletz 1999), Fault Tree Analysis (FTA) and Event Tree Analysis (ETA) (Storey 1996), and more recent methods such as Functional Resonance Analysis Method (FRAM) (Hollnagel 2012) and System Theoretic Process Analysis (STPA) (Leveson 2012). The reader is directed to these references for in-depth descriptions of the methods.

Risks are often assessed qualitatively first, informed by engineering judgement and gut feelings. This type of qualitative analysis is considered sufficient in many cases (Health and Safety Executive 2001). In situations where this qualitative analysis suggests that risks might be high or the consequences severe, quantitative analysis might be carried out. Quantitative analysis is more common for engineering problems than for organisational changes because such changes are more difficult to model, and the uncertainties associated with quantitative estimates can be very high.

Common to all of the different approaches for risk assessment is the proactive search for threats and vulnerabilities. In proactive approaches, the emphasis shifts from historic event and outcome data (i.e. adverse events) towards consideration of future events (i.e. risk). Regulatory requirements [e.g. (Health and Safety Executive 2001)] act as a strong motivator, but there are also ethical and societal considerations. In addition, organisations are increasingly aware of the potentially negative impact of poor safety performance on the reputation of a business or company (Sujan et al. 2017).

2.5 Risk Management

Managing risk is more than describing and assessing risks. This is often not well understood. One might be tempted to equate the assessment of risk with decision-making about risk. However, this latter process is much more complex than simply reviewing the facts and figures produced by the (technical) risk assessment. Managers have to make decisions about whether or not risks are acceptable, and whether to invest money in order to reduce risk. Crucially, as the risk management process proceeds from risk analysis towards risk-informed decision-making, judgements about risk tend to be less based on purely factual evidence and more based on value assessments instead (Aven 2016). Value-based judgements might include considerations other than safety, such as production benefits and other business impacts, ethical concerns, issues of corporate responsibility, and whether or not a decision might hold up in court (Sujan et al. 2017). Social and cultural theories of risk, such as Social Amplification of Risk (see above), can help to explain how decisions about risks are made.

In the UK context, safety-critical industries follow the framework set out in the Health and Safety at Work etc. Act 1974 (1974) and the guidance developed by the HSE (Health and Safety Executive 2001). A central principle of the Health and Safety at Work Act is that those who create risks are responsible for protecting workers and the public from the consequences. A key notion is the concept of 'reasonable practicability', used to demonstrate that risks have been controlled effectively. Reasonable practicability was first defined by the Court of Appeal in its judgement in Edwards versus National Coal Board in 1949. This case established that operators of systems have a legal duty to reduce risk unless the sacrifices (in terms of money, effort, etc.) are grossly disproportionate to the expected benefits. This principle is known as *So far as is reasonably practicable* (SFAIRP) or *As Low As Reasonably Practicable* (ALARP) in its practical application. It requires from the operator a conscious and transparent decision about whether or not risk control measures are put in place.

In current practice, the risk space is divided into three different regions: the region of unacceptable risk, where societal concerns are so great that the system cannot be operated; the region of negligible risk, where the risk is perceived to be so small that no further action is required to mitigate the risk; and the region of tolerable risk—this is the region where the risk is perceived to be tolerable, but only if further risk reduction is impracticable or if the associated costs are grossly disproportionate.

Reducing a risk to be ALARP involves weighing the risk against the sacrifice (whether in terms of money, time or trouble) needed to further reduce it. The presumption of ALARP is that available risk reduction measures will be put in place and risk will continue to be reduced until the situation where the cost (sacrifice) of further risk reduction is grossly disproportionate to the benefits that would be achieved. ALARP is therefore not simply a process of balancing the cost and benefits of risk reduction measures. It is a principle that compels risk reduction until the point that further risk reduction can be (justifiably) ruled out.

Affordability of risk reduction is not a consideration in ALARP judgements as such, i.e. it is not acceptable (under the ALARP principle) to justify the non-implementation of a risk reduction measure simply due to the lack of resources for implementation. The ALARP approach recognises that absolute safety might not be achievable, and it allows duty holders to take a proportionate approach to the management of risk.

In many situations, the justification for ALARP can be made qualitatively and with reference to existing 'good practice'. Good practice refers to those principles, practices and risk control measures that the regulator would typically expect to see in comparable situations. Good practice guidance is established through a continuous-consensus-based process with a range of stakeholders, including employers, trade unions and manufacturers.

There may be cases, however, where there is no established good practice (e.g. novel technology or systems), or where the situation is complex and a decision cannot be reached easily based on reference to good practice alone. In situations, where the qualitative judgement is difficult, a quantitative Cost-Benefit Analysis (CBA) might provide additional input into the decision-making process. This is often the case in situations where the risk is close to the intolerable range.

In CBA, both the sacrifice (investments, effort, training, maintenance, etc.) and the benefits (risk reduction) have to be expressed in monetary terms. Then, a judgement about gross disproportion can be made using a suitably chosen disproportion factor. The disproportion factor indicates by what factor the costs need to outweigh the expected benefits in order to be able to claim gross disproportionality. The HSE suggests a rule of thumb for selecting the disproportion factor: a factor of 3 for risks to workers (i.e. costs that are three times higher than benefits), a factor of 2 for low risks to the public, and a factor of 10 for significant risks to the public.

The quantification of benefits requires that an economic value be put on human life and suffering. The Department for Transport provides and updates an estimated value of preventing a statistical fatality (VPF), which is also commonly adopted or referred to in other industries. The VPF is derived from willingness-to-pay studies and does not indicate that a value is placed on an individual life. The intention of VPF is to express what people are prepared to pay to secure a certain averaged risk reduction. For example, a VPF of £1,000,000 corresponds to a reduction in risk of 1/100,000 being worth £10 to an average individual (Health and Safety Executive 2001).

While the concept of ALARP is widely accepted in the UK, its application has not been without problems and controversy. For example, the Nimrod review into the loss of a Royal Air Force aircraft in Afghanistan in 2006 (Haddon-Cave 2009) extensively dissects the ALARP judgements made concerning risks by the Nimrod Integrated Project Team, suppliers, and independent advisors, and questions whether the principle of ALARP was fully understood. The HSE recognises that deciding whether a risk is reduced ALARP can be challenging and that the decision relies on expert judgement (Health and Safety Executive 2001).

Regulatory guidance across the industries emphasises that quantitative CBA cannot be relied upon as the only input into the ALARP decision. While CBA calculations have the appeal of numerical precision and perceived objectivity, the underlying mathematical models can be complex, they might rely on estimates of low confidence, and they might make significant simplifications. For example, in the nuclear industry, elaborate models have been proposed to assess the economic impact of nuclear accidents, including both immediate and indirect consequences that result from the impact on tourism and businesses, and wider health impact costs (Higgins et al. 2008).

A further problem with ALARP and CBA is that each duty holder manages safety-cost trade-offs separately, but the safety measures might have an impact on the system as a whole. For example, the aviation system consists of different stakeholders, such as airlines, aircraft manufacturers, equipment suppliers, maintenance companies, airports and air traffic service providers. Together, these stakeholders aim to provide safe air travel. Lack of investments in winter preparedness (de-icing procedures, etc.) at an airport, for example, might not have a direct significant impact on safety at the airport, but might affect the airspace and other airports in case flights need to be diverted in large numbers.

In addition to these technical challenges to the ALARP concept, there exists an argument questioning the ethics of ALARP both within the UK as well as in other countries. Such an opposing position calls for the implementation of safety measures capable of eliminating risk regardless of cost. If this is not the case, then it is implied that a preventable accident is acceptable to organisations, the regulators and the government. This ethical argument is particularly powerful following accidents in which the victims had no immediate involvement, such as the population in the proximity of a site where toxic materials were released. However, the problem with this argument is that it offers no account of the size of the benefits or measure of the costs, and how these compare to one another.

2.6 Patient Safety Risk Management

The management of risks to patient safety is still predominantly reactive (Sujan 2015). Common tools for risk management include root cause analysis (RCA) and incident reporting. These approaches look at adverse events and incidents, trying to identify factors that contributed specifically to these events, so that remedial action can be undertaken. While useful information can be generated in this way, the downside is that these approaches are reactive, i.e. they usually only look at events that have already caused harm.

Increasingly, organisations are encouraged to adopt proactive risk management approaches. FMEA has been proposed frequently as a potential tool for use in healthcare contexts. In particular, the Veterans' Affairs (VA) in the USA has been promoting this approach, and a healthcare-specific version has been developed—Healthcare Failure Mode and Effects Analysis (DeRosier et al. 2002). FMEA has been used in different settings including blood transfusion and emergency care (Burgmeier 2002; Sujan and Felici 2012). Often participants have described the experiences of using FMEA positively, but there has been a criticism of the approach (Dean Franklin et al. 2012; Shebl et al. 2009). It has been suggested that FMEA is unduly time-consuming and that the risk assessments produced using FMEA were dependent on the participants and not necessarily replicable. One can add to these criticisms that knowledge of FMEA and other proactive methods for risk analysis is still limited in many healthcare organisations, and that such approaches are used only infrequently.

2.7 Safety Cases—Demonstrating and Critiquing the Safety Position

The Concept of a Safety Case

The HSE in the UK requires that manufacturers and operators of safety-critical systems demonstrate that they have adopted a thorough and systematic process for understanding proactively the risks associated with their systems, and that these risks have been controlled appropriately. The regulator does not specify how risks should be dealt with specifically, but rather sets goals that have to be achieved (e.g. all risks have to be reduced ALARP). It is then left to the manufacturers and operators of systems to argue that they have met these goals. The benefit of this goal-based approach over a purely prescriptive approach is the flexibility it provides. Prescriptive approaches are based on past experience and are best suited for well-established systems. However, as systems change, and with the introduction of new technologies, prescriptive approaches quickly become outdated, they might hinder innovation, and they might even affect safety adversely. The goal-based approach, on the other hand, provides the necessary flexibility that allows new risks to be addressed, or to enable operators to use new ways of controlling risks. In practice, regulation typically involves a mixture of both approaches.

In the UK, these duties are often fulfilled through the use of safety cases (Maguire 2006). The purpose of a safety case can be described as providing a structured argument, supported by a body of evidence that provides a compelling, comprehensible and valid case that a system is acceptably safe for a given application in a given context (UK Ministry of Defence 2007). A key component of any safety case is a risk-based argument and corresponding evidence. This is intended to demonstrate that all risks associated with a particular system have been identified, that appropriate risk controls have been put in place, and that there are appropriate processes in place to monitor the effectiveness of the risk controls and the safety performance of the system on an on-going basis. The argument and evidence in safety cases are then examined and challenged, typically by independent safety assessors, as part of the overall safety assessment or certification process.

As mentioned above, the safety case can usefully document assumptions and uncertainties, which go beyond the technical (and numerical) assessment of risk. Increasingly, best practice is to include a 'confidence argument' to complement the risk-based argument (Hawkins et al. 2011). This confidence argument outlines the strength of the evidence and the extent to which one can place confidence in the safety case. In practice, safety case assessors tend to challenge issues of qualitative nature (i.e. assumptions, boundaries of the system, excluded scenarios, etc.) rather than specific numerical values.

As mentioned above regarding the case of the Nimrod accident, safety cases are not without criticism (Leveson 2011; Steinzor 2011). These criticisms are directed at the practice of using safety cases, not necessarily at the concept itself (though it might be hard to distinguish between the two). Risk management activities might be geared more towards producing 'paper safety' as documented in the safety case rather than at achieving real improvements in safety, and safety cases might not be updated or maintained even though systems continuously evolve. However, the fundamental principles of a safety case, namely the thorough and proactive consideration of risk, the openness and transparency, and the external scrutiny and critique are generally accepted as best practice.

2.8 Using Safety Cases in Healthcare

There has been some recent interest in the application of safety cases in healthcare (Health Foundation 2014; Sujan et al. 2013, 2015b). However, at present, this interest is mostly limited to medical devices and health informatics applications. In the USA, the FDA has issued guidance for assurance cases (a type of safety case) for infusion pumps that are certified via the 510(k) route (FDA 2014). In the UK, NHS Digital has published standards for both manufacturers and users of health IT products, which include a requirement for the development of a clinical safety case (Health and Social Care Information Centre 2016a, b). There is very limited empirical evidence available about the use of safety cases in healthcare, in particular at the system or the service level (Sujan et al. 2015b).

The regulatory environment acts as a key driver for the use of safety cases in the UK industries (Sujan et al. 2017). In the NHS, and probably in healthcare more generally, there is no single body to provide centralised and coordinated oversight of patient safety. There are around 20 regulatory bodies in health and social care in England, and this diversity has contributed to a lack of a coherent push for improving patient safety. In addition, regulatory bodies require the necessary technical understanding and adequate resources in order to make a safety case approach work in practice.

In the absence of a regulatory push for safety cases, organisations might still consider using safety cases to make their risk position explicit. This requires adaptations to the safety case concept, moving it from a regulatory tool towards a tool for effective risk management. This is in line with observations and suggestions by both the Cullen inquiry (following the Piper Alpha oil platform explosion) (The Honourable Lord Cullen 1990) and the Haddon-Cave report (following the loss of a Royal Air Force aircraft in Afghanistan) (Haddon-Cave 2009). Lord Cullen's report argues that safety cases should first and foremost provide assurance to companies themselves that they have followed a systematic and thorough approach to risk management to ensure that their systems are safe. Similarly, while the Haddon-Cave report criticises the Ministry of Defence and BAE Systems for their safety culture and attitudes, the report suggests that safety cases remain central to making an organisation's risk position explicit so that it can be reviewed and critiqued. A practical account of the use of Safety Cases is given in Section II, Chap. 6.

2.9 Organisational Learning

The third safety management practice, which successful organisations pursue, is organisational learning. Organisational learning can be characterised as a continuous cycle of action and reflection (Carroll and Edmondson 2002). Organisations might be more successful at learning from past experience if they create and foster the capacity for deep reflection on whole system dynamics, which can lead to fundamental change (Argyris and Schön 1996). On the other hand, insistence on past traditions, and quick fixes to existing strategies and procedures might inhibit more powerful forms of organisational learning. Organisations have a range of learning processes at their disposal, which might be internal (for example, audits and adverse event reviews) as well as external (for example, feedback from the regulator) (Popper and Lipshitz 1998).

Many organisations are relying on incident-reporting systems as a key process for reporting and organisational learning (Drupsteen and Guldenmund 2014; Le Coze 2013; Lukic et al. 2010). Ideally, effective learning from incidents triggers improvements in practice that enhance safety and productivity (Lukic et al. 2012). The analysis of incidents seeks to reveal contributory factors and underlying causes (Drupsteen and Guldenmund 2014), which can then be addressed in order to reduce the likelihood of incidents recurring. Learning from past experiences does not have to be limited to the consideration of incidents, but can also include monitoring and analysis of leading indicators, or even weak signals (Drupsteen and Wybo 2015). However, there is increasing evidence in the literature that suggests that effective learning from past experiences in order to improve safety performance remains challenging even in traditional safety-critical industries (Le Coze 2013; Lukic et al. 2012; Drupsteen and Hasle 2014).

2.10 The Challenges of Organisational Learning in Healthcare

Following the public inquiry into the failings at Mid Staffordshire NHS Foundation Trust, the subsequent Berwick report generated lessons and suggestions for change for the UK government and the National Health Service (NHS) in England (National Advisory Group on the Safety of Patients in England 2013). The report recommends that the NHS should aim to become a system devoted to continuous learning and improvement of patient care. This is clearly a fundamental requirement for any healthcare organisation aspiring to improve the safety of care to higher levels.

Incident reporting as an instrument for organisational learning was introduced into the NHS about 2003, following a publication by the Department of Health (2000). This report recommended the development of a reporting system based on the model of incident reporting used in commercial aviation. Incident reporting is well established in the NHS, and it is regarded as a key instrument for improving patient safety and the quality of services (Anderson et al. 2013; Barach and Small 2000).

In one respect, incident reporting in the NHS has been very successful. There are a staggering number of incidents reported every year. However, despite a large number of potential learning opportunities, questions have been raised about the effectiveness of incident-reporting systems to contribute to improvements in patient safety (Pasquini et al. 2011; Sujan and Furniss 2015; Braithwaite et al. 2010; Macrae 2015; Vincent 2004). There are now many studies that document barriers to effective incident reporting in healthcare. Such barriers include, for example, fear of blame and repercussions, poor usability of incident-reporting systems, perceptions among doctors that incident reporting is a nursing process, lack of feedback to staff who report incidents, and lack of visible improvements to the local work environment as a result of reported incidents (Benn et al. 2009; Braithwaite et al. 2010; Lawton and Parker 2002; Macrae 2015; Sujan 2012; Sujan et al. 2011). Among management staff, in particular, there continues to be a widespread misperception that incident-reporting systems might be useful for monitoring incident frequencies, despite evidence that suggests that incident-reporting data are poor indicators of actual incident frequencies (Westbrook et al. 2015). It has been suggested that the focus of learning from incidents in healthcare has been too much on collecting and categorising data (Macrae 2015; Anderson and Kodate 2015), whereas successful learning from experience should inherently be a social and participative process (Macrae 2015; Lukic et al. 2012).

2.11 Learning from the Ordinary

How can healthcare organisations enhance their ability to learn from past experience in order to set them on the path towards becoming ultra-safe organisations given the obstacles and practical difficulties with learning from incidents outlined above? One way might be to shift the focus from formal learning about extraordinary failures and incidents towards more de-centralised, local forms of learning about everyday clinical work (Sujan 2015, 2018).

An example of such a local form of learning is the Proactive Risk Monitoring in Healthcare (PRIMO) approach. This approach to organisational learning was developed in order to elicit a rich contextual picture of the local work environment, to move away from negative and threatening notions of errors and mistakes, and to encourage active participation and ownership with clear feedback for local work practices (Sujan 2012; Sujan et al. 2011). The distinguishing feature of the PRIMO approach is that it focuses on learning from the ordinary; in this case, the various hassles that practitioners experience in their everyday clinical work.

Hassle, in this instance, can be defined loosely as anything that causes people problems during their daily work. Examples of hassle include, for instance, unavailable equipment such as drip stands on a ward or supporting equipment for undertaking radiographic procedures. There are a number of important benefits of learning from everyday hassle. Among these, the most important benefit is arguable that the focus on hassle supports building an understanding of the system dynamics, i.e. of the way performance adjustments are made, and the way work ordinarily unfolds. Reports of hassle typically contain not only descriptions of how the hassle manifested itself, but also how people coped—how they adapted their behaviour in order to continue to provide safe and good quality care (Sujan et al. 2015a). Examples of typical adaptations made by healthcare professionals include the sharing of information and personal negotiation to create a shared awareness, prioritisation of goals and of activities, and offering and seeking help.

Other local and informal processes that organisations might consider supporting include regular staff meetings aimed at identifying ways to improve the delivery of care, informal discussions between staff and their managers, and discussions among peers, and informal lunchtime improvement groups. Such processes are perceived as locally owned, and they might be better suited to provide shared awareness, to make staff feel that they are being listened to and that they can make a contribution to improving patient safety, and for generating ownership for improvement interventions (Sujan 2015).

Research suggests that where organisational effort is invested to support and include such processes, these can have a positive effect on staff engagement in reporting and learning activities (Sujan 2012) and on patient safety (Goldenhar et al. 2013). Utilising a range of processes that draw upon and strengthen different aspects of an organisation's culture might enable healthcare organisations to deliver more sustainable improvements in patient safety (Singer and Vogus 2013).

2.12 Is Healthcare a Safety-Critical Industry?

The publication of reports in the USA and in the UK around the year 2000 about the significant extent of harm to patients triggered policy initiatives and changes with the aim of improving patient safety (Department of Health 2000; Kohn et al. 2000). From the very start, policy makers and researchers looked at practices from other industries to inform improvement efforts in healthcare. For example, one of the best-known initiatives is the adoption of incident reporting, which was modelled on the success of the Aviation Safety Reporting System operated by NASA (Billings 1998). However, as the 'frustrating case of incident-reporting systems' (Shojania 2008) in healthcare has shown, the transfer of lessons from one setting to another is not unproblematic. Parallels have been drawn between healthcare and other industries, such as aviation, but the fact is that while there are similarities to a certain extent, healthcare is a complex system with its own characteristics and idiosyncrasies. Healthcare is an incredibly diverse domain that requires frequent interaction across organisational boundaries, and that demands a significant tolerance for increased levels of uncertainty (Lyons et al. 2004; Sujan et al. 2015c). Owing to these differences in the organisational, institutional and cultural context, methods and techniques from other industries have to be applied with caution and have to be adapted appropriately (Sujan et al. 2016, 2017). We have to understand the purpose, theoretical underpinnings and the limitations of these techniques within their original context, in order to apply them meaningfully within healthcare (Sutcliffe et al. 2017). Failure to do so might limit the benefits we can expect to get from industrial safety management approaches imported into healthcare, and it might even contribute to increasing risks to patients (Clay-Williams and Colligan 2015).

In this chapter, we have outlined the rationale behind the adoption of key safety management practices. It is instructive to conclude the chapter with a look at some of the key challenges that need to be addressed when adopting these practices in a healthcare context. These challenges are role of the patient, the relatively low levels of reliability of many clinical processes, and the different regulatory focus (Sujan et al. 2018).

2.13 Patient Perception of Risk

Patients have an active part in their own care. This is in contrast to traditional safety-critical industries, where the public are assumed to be passive consumers. Healthcare services have been described as being co-produced by patients and healthcare providers (Batalden et al. 2016). While this might still be regarded as an ideal, it has important implications for how we approach the concept of risk, which is at the heart of industrial safety management practices. The ISO definition of risk introduced above regards risk as 'effect of uncertainty on objectives'. In the context of healthcare, we might ask on whose objectives—the clinician's or the patient's objectives? Are these necessarily the same objectives? And are the effects of uncertainty on these objectives the same? Considering the extreme, but frequent example of patient death—from the clinician's perspective death might be considered a discrete outcome with a certain severity. However, from the patient's perspective, the nature of death might make all the difference (Wears 2014). Am I allowed to die peacefully at home, or will death be painful and in a clinical setting? These are fundamental questions for patients. Technical assessment of risk might inform the advice given by clinicians, but patients might consider wider, and more personal aspects in their decision-making about risk, which are not considered in the technical assessment. Bob Wears suggested that in a healthcare context, risk might be better understood as something personal that is negotiated between the patient and clinicians (Wears 2014).

2.14 Reliability of Clinical Processes

Assessment of risk presupposes a good understanding of the system and the processes. In safety-critical industries, many processes are highly standardised and have very high reliability. This is the starting point for risk analysis, which is often concerned with high-severity, low-probability scenarios, i.e. with very infrequent events that might have catastrophic consequences. In healthcare, this situation is markedly different, because clinical processes and tasks often have very poor reliability. Many processes such as handover occur regularly but have never been designed, and hence exhibit a lot of variabilities depending on the individuals involved (Sujan et al. 2015a). The absence of properly designed processes is almost a given in situations where care crosses departmental and organisational boundaries.

Studies of different clinical processes found that the reliability of many care processes is very poor when compared with other safety-critical industries. For example, a US study found that only 55% of patients received care that was consistent with best practice, with reliability figures ranging from 79% (senile cataract) to as little as 10% (alcohol dependence) (McGlynn et al. 2003). More recently, a large study investigating the quality of care provided to children in Australia found that across a range of conditions adherence to recommended quality indicators was only about 60% (Braithwaite et al. 2018). The authors caution that poor adherence to guidelines, for example, asthma guidelines may negatively affect patient outcomes. Patient safety interventions, such as the World Health Organisation (WHO) surgical checklist (Haynes et al. 2009) are troubled by persistently high levels of non-adherence (Vats et al. 2010; Pickering et al. 2013).

The poor reliability of clinical processes, and the absence of documented tasks and process models pose challenges for the transfer of lessons from other industries. One could even question whether it makes much sense to apply methods that are concerned with the occasional catastrophe to situations of everyday disaster in healthcare, i.e. to scenarios that have high frequency and minor or moderate severity of their consequences. In some respects, the risks in many healthcare processes are known, but have not been addressed.

However, another way of approaching this challenge more broadly is to question whether the absence of process and task models is a desirable or necessary state. The systematic application of, for example, risk analysis approaches could contribute to raising awareness about the unsatisfactory status quo, and support clinicians and healthcare providers in attempts to define and standardise clinical processes. The safety case could become a communication tool that allows clinical teams to document their risks and to communicate these to others, e.g. across departmental boundaries or across the organisational hierarchy, where solutions might be found and resources allocated.

2.15 The Focus of Regulation

The development of the concepts and approaches described in this chapter has been tightly linked to strong, proactive regulatory regimes. These are typically risk-based and entail principles such as ALARP (described above). This is in stark contrast to the current regulatory situation in healthcare. There is a lack of regulatory guidance and an absence of institutional drivers for systematically and proactively reducing risk (Sujan et al. 2017). Healthcare providers have few regulatory incentives for investing time and resource in the use of systematic safety management processes as is done in other industries, and—more importantly—there is a critical lack of awareness of and familiarity with the application of relevant techniques and practices among healthcare providers, healthcare professionals and the regulatory bodies.

It is highly unlikely that the regulatory regime in healthcare will look anything like those in other safety-critical industries in the short-to-medium term. The introduction of a risk-based approach requires a different mindset, which can be met with scepticism. The techniques can be perceived as overly technical, too resource-intensive, and too variable in their outputs (Dean Franklin et al. 2012; Shebl et al. 2009). In addition, at present, there is a lack of rigorous evaluation of the application of risk assessment and risk management techniques in terms of their impact on patient safety. This is crucial, because healthcare is governed by strong credence in the principle of evidence-based medicine. It is imperative, therefore, to give careful attention to the norms, values and needs of the different stakeholders by addressing, for example, issues around the existing knowledge base, the role of scientific evidence in healthcare, and the need of healthcare providers to demonstrate accountability in prescribed ways to regulators, commissioners and patients (Dixon-Woods et al. 2014).

2.16 Summary

This chapter discussed three key strategies that successful safety-critical industries adopt in order to achieve outstanding safety performance: the proactive management of risk, the explicit demonstration and critique of the organisation's safety position, and a commitment to continuous learning and improvement.

In principle, these strategies can work across different industries, and they have the potential to transform radically the safety record of healthcare organisations. However, healthcare is unlike other safety-critical industries in many aspects, and the different cultural and contextual backgrounds have to be considered. None the less, these lessons from industry should provide valuable input to patient safety management efforts in healthcare.

References

Althaus, C. E. (2005). A disciplinary perspective on the epistemological status of risk. *Risk Analysis, 25,* 567–588.

Amalberti, R., Auroy, Y., Berwick, D., & Barach, P. (2005). Five system barriers to achieving ultrasafe health care. *Annals of Internal Medicine, 142,* 756–764.

Anderson, J. E., & Kodate, N. (2015). Learning from patient safety incidents in incident review meetings: Organisational factors and indicators of analytic process effectiveness. *Safety Science, 80,* 105–114.

Anderson, J. E., Kodate, N., Walters, R., & Dodds, A. (2013). Can incident reporting improve safety? Healthcare practitioners' views of the effectiveness of incident reporting. *International Journal for Quality in Health Care, 25,* 141–150.

Argyris, C., & Schön, D. A. (1996). *Organisational learning II: Theory, method and practice.* Reading, MA: Addison-Wesley.

Aven, T. (2011). On the new ISO guide on risk management terminology. *Reliability Engineering & System Safety, 96,* 719–726.

Aven, T. (2012). The risk concept—Historical and recent development trends. *Reliability Engineering & System Safety, 99,* 33–44.

Aven, T. (2016). Risk assessment and risk management: Review of recent advances on their foundation. *European Journal of Operational Research, 253,* 1–13.

Barach, P., & Small, S. D. (2000). Reporting and preventing medical mishaps: lessons from non-medical near miss reporting systems. *BMJ, 320,* 759–763.

Batalden, M., Batalden, P., Margolis, P., Seid, M., Armstrong, G., Opipari-Arrigan, L., et al. (2016). Coproduction of healthcare service. *BMJ Quality & Safety, 25,* 509–517.

Benn, J., Koutantji, M., Wallace, L., Spurgeon, P., Rejman, M., Healey, A., et al. (2009). Feedback from incident reporting: information and action to improve patient safety. *Quality and Safety in Health Care, 18,* 11–21.

Billings, C. (1998). Incident reporting systems in medicine and experience with the aviation reporting system. In R. I. Cook, D. Woods, & C. A. Miller (Eds.), *A tale of two stories: contrasting views of patient safety*. US National Patient Safety Foundation: North Adams, MA.

Braithwaite, J., Hibbert, P. D., Jaffe, A., et al. (2018). Quality of health care for children in Australia, 2012–2013. *JAMA, 319,* 1113–1124.

Braithwaite, J., Westbrook, M. T., Travaglia, J. F., & Hughes, C. (2010). Cultural and associated enablers of, and barriers to, adverse incident reporting. *Quality and Safety in Health Care, 19,* 229–233.

Brenkert-Smith, H., Dickinson, K. L., Champ, P. A., & Flores, N. (2013). Social amplification of wildfire risk: The Role of social interactions and information sources. *Risk Analysis: An International Journal, 33,* 800–817.

Burgmeier, J. (2002). Failure mode and effect analysis: An application in reducing risk in blood transfusion. *Joint Commission Journal on Quality Improvement, 28,* 331–339.

Carroll, J. S., & Edmondson, A. C. (2002). Leading organisational learning in health care. *Quality And Safety In Health Care, 11,* 51–56.

Clay-Williams, R., & Colligan, L. (2015). Back to basics: Checklists in aviation and healthcare. *BMJ Quality & Safety, 24,* 428–431.

Dean Franklin, B., Shebl, N. A., & Barber, N. (2012). Failure mode and effects analysis: Too little for too much? *BMJ Qual Saf, 21,* 607–611.

Department of Health. (2000). *An organisation with a memory*. London: The Stationery Office.

Derosier, J., Stalhandske, E., Bagian, J. P., & Nudell, T. (2002). Using health care failure mode and effect analysis: The VA National center for patient safety's prospective risk analysis system. *Joint Commission Journal on Quality Improvement, 28*(248–67), 209.

Dixon-Woods, M., Martin, G., Tarrant, C., Bion, J., Goeschel, C., Pronovost, P., et al. (2014). *Safer clinical systems: Evaluation findings*. London: Health Foundation.

Drupsteen, L., & Guldenmund, F. W. (2014). What is learning? A review of the safety literature to define learning from incidents, accidents and disasters. *Journal of Contingencies and Crisis Management, 22,* 81–96.

Drupsteen, L., & Hasle, P. (2014). Why do organizations not learn from incidents? Bottlenecks, causes and conditions for a failure to effectively learn. *Accident Analysis and Prevention, 72,* 351–358.

Drupsteen, L., & Wybo, J.-L. (2015). Assessing propensity to learn from safety-related events. *Safety Science, 71*(Part A), 28–38.

FDA. (2014). *infusion pumps total product life cycle: Guidance for industry and FDA staff.* MD: Rockville.

Goldenhar, L. M., Brady, P. W., Sutcliffe, K. M., & Muething, S. E. (2013). Huddling for high reliability and situation awareness. *BMJ Quality & Safety, 22,* 899–906.

Haddon-Cave, C. (2009). *The Nimrod review: An independent review into the broader issues surrounding the loss of the RAF Nimrod MR2 aircraft XV230 in Afghanistan in 2006*. London: The Stationary Office.

Hawkins, R., Kelly, T., Knight, J., & Graydon, P. (2011). A new approach to creating clear safety arguments. In C. Dale & T. Anderson (Eds.), *advances in systems safety*. London: Springer.

Haynes, A. B., Weiser, T. G., Berry, W. R., Lipsitz, S. R., Breizat, A.-H. S., Dellinger, E. P., et al. (2009). A surgical safety checklist to reduce morbidity and mortality in a global population. *New England Journal of Medicine, 360,* 491–499.

Health and Safety at Work etc. Act (1974). Available http://www.legislation.gov.uk/ukpga/1974/37. Accessed May 24, 2019.

Health & Social Care Information Centre. (2016a). Clinical risk management: Its application in the deployment and use of health IT systems—Implementation guidance (SCCI 0160) [Online]. Available http://content.digital.nhs.uk/media/20988/0160382012spec/pdf/0160382012spec.pdf. Accessed March 16, 2017.

Health & Social Care Information Centre. (2016b). Clinical risk management: Its application in the manufacture of health IT systems—implementation guidance (SCCI 0129) [Online]. Available: http://content.digital.nhs.uk/media/20984/0129392012spec/pdf/0129392012spec.pdf. Accessed March 16, 2017.

Health and Safety Executive. (2001). *Reducing risk: Protecting people*, Norwich, Her Majesty's Stationary Office.

Health Foundation. (2014). *Exploring the potential use of safety cases in health care.* London: Health Foundation.

Higgins, N.A., Jones, C., Munday, M., Balmforth, H., Holmes, W., Pfuderer, S., Mountford, L., Harvey, M. & Charnock, T. (2008). *COCO-2: A model to assess the economic impact of an accident.* London: Health Protection Agency.

Hollnagel, E. (2012). *Fram, the functional resonance analysis method: Modelling complex socio-technical systems.* Ashgate Publishing Ltd.

Health and Safety Executive. ALARP "at a glance" [Online]. Available: https://www.se.gov.uk/risk/theory/alarpglance.htm. Accessed May 24, 2019.

IEC. (2006). *Analysis techniques for system reliability—Procedure for failure mode and effects analysis (FMEA).* Geneva: International Electrotechnical Commission. (IEC 60812 Ed2.0).

Kapur, N., Parand, A., Soukup, T., Reader, T., & Sevdalis, N. (2016). Aviation and healthcare: A comparative review with implications for patient safety. *JRSM Open, 7.* https://doi.org/10.1177/2054270415616548.

Kasperson, R. E., Renn, O., Brown, H. S., Emel, J., Goble, R., Kasperson, J. X., et al. (1988). The social amplification of risk: A conceptual framework. *Risk Analysis, 8,* 177–187.

Kletz, T. A. (1999). *Hazop And Hazan: Identifying and assessing process industry hazards.* Institution Of Chemical Engineers: Rugby.

Kohn, L. T., Corrigan, J. M., & Donaldson, M. S. (2000). *To err is human: Building a safer health system.* Washington: The National Academies Press.

Lawton, R., & Parker, D. (2002). Barriers to incident reporting in a healthcare system. *Quality and Safety in Health Care, 11,* 15–18.

Le Coze, J. C. (2013). What have we learned about learning from accidents? Post-disasters reflections. *Safety Science, 51,* 441–453.

Leveson, N. (2011). The use of safety cases in certification and regulation. *Journal of System Safety, 47.*

Leveson, N. (2012). *Engineering a safer world.* Cambridge, MA: MIT Press.

Lukic, D., Littlejohn, A., & Margaryan, A. (2012). A framework for learning from incidents in the workplace. *Safety Science, 50,* 950–957.

Lukic, D., Margaryan, A., & Littlejohn, A. (2010). How organisations learn from safety incidents: A multifaceted problem. *Journal Of Workplace Learning, 22,* 428–450.

Lyons, M., Adams, S., Woloshynowych, M., & Vincent, C. (2004). Human reliability analysis in healthcare: A review of techniques. *International Journal Of Risk & Safety In Medicine, 16,* 223–237.

Macrae, C. (2015). The problem with incident reporting. *BMJ Quality & Safety, 25,* 71–75.

Maguire, R. (2006). *Safety cases and safety reports.* Aldershot: Ashgate.

Mcglynn, E. A., Asch, S. M., Adams, J., Keesey, J., Hicks, J., Decristofaro, A., et al. (2003). The quality of health care delivered to adults in the United States. *New England Journal of Medicine, 348,* 2635–2645.

National Advisory Group on the Safety of Patients in England. (2013). *A promise to learn—A commitment to act*. London: Department of Health.

Pasquini, A., Pozzi, S., Save, L., & Sujan, M. A. (2011). Requisites for successful incident reporting in resilient organisations. In E. Hollnagel, J. Paries, D. Woods, & J. Wreathall (Eds.), *Resilience engineering in practice: A guidebook*. Farnham: Ashgate.

Pickering, S. P., Robertson, E. R., Griffin, D., Hadi, M., Morgan, L. J., Catchpole, K. C., et al. (2013). Compliance and use of the World Health Organization checklist in uk operating theatres. *British Journal of Surgery, 100,* 1664–1670.

Popper, M., & Lipshitz, R. (1998). Organizational learning mechanisms a structural and cultural approach to organizational learning. *The Journal Of Applied Behavioral Science, 34,* 161–179.

Shebl, N. A., Franklin, B. D., & Barber, N. (2009). Is failure mode and effect analysis reliable? *Journal of Patient Safety, 5.*

Shojania, K. G. (2008). The frustrating case of incident-reporting systems. *Quality and Safety in Health Care, 17,* 400–402.

Singer, S. J., & Vogus, T. J. (2013). Reducing hospital errors: Interventions That build safety culture. *Annual Review of Public Health, 34,* 376–396.

Steinzor, R. (2011). Lessons from the North Sea: Should "safety cases" come to America? *Boston College Environmental Affairs Law Review, 38,* 417–444.

Storey, N. (1996). *Safety-critical computer systems*. Harlow: Pearson Prentice Hall.

Sujan, M. (2015). An organisation without a memory: A qualitative study of hospital staff perceptions on reporting and organisational learning for patient safety. *Reliability Engineering & System Safety, 144,* 45–52.

Sujan, M. (2018). A safety-II perspective on organisational learning in healthcare organisations; Comment on "false dawns and new horizons in patient safety research and practice". *International Journal of Health Policy And Management, 7,* 662–666.

Sujan, M., & Furniss, D. (2015). Organisational Reporting and learning systems: Innovating inside and outside of the box. *Clinical Risk, 21,* 7–12.

Sujan, M., Spurgeon, P., & Cooke, M. (2015a). The role of dynamic trade-offs in creating safety—A qualitative study of handover across care boundaries in emergency care. *Reliability Engineering & System Safety, 141,* 54–62.

Sujan, M., Spurgeon, P., Cooke, M., Weale, A., Debenham, P., & Cross, S. (2015b). The development of safety cases for healthcare services: Practical Experiences, opportunities and challenges. *Reliability Engineering & System Safety, 140,* 200–207.

Sujan, M. A. (2012). A novel tool for organisational learning and its impact on safety culture in a hospital dispensary. *Reliability Engineering & System Safety, 101,* 21–34.

Sujan, M. A., Chessum, P., Rudd, M., Fitton, L., Inada-Kim, M., Cooke, M. W., et al. (2015c). Managing competing organizational priorities in clinical handover across organizational boundaries. *Journal Of Health Services Research & Policy, 20,* 17–25.

Sujan, M. A., Embrey, D., & Huang, H. (2018). On the application of human reliability analysis in healthcare: Opportunities and challenges. *Reliability Engineering & System Safety*. https://doi.org/10.1016/j.ress.2018.06.017.

Sujan, M. A., & Felici, M. (2012). Combining failure mode and functional resonance analyses in healthcare settings. *Computer Safety, Reliability, and Security* 364–375.

Sujan, M. A., Habli, I., Kelly, T. P., Gühnemann, A., Pozzi, S., & Johnson, C. W. (2017). How can health care organisations make and justify decisions about risk reduction? Lessons from a cross-industry review and a health care stakeholder consensus development process. *Reliability Engineering & System Safety, 161,* 1–11.

Sujan, M. A., Habli, I., Kelly, T. P., Pozzi, S., & Johnson, C. W. (2016). Should healthcare providers do safety cases? Lessons from a cross-industry review of safety case practices. *Safety Science, 84,* 181–189.

Sujan, M. A., Huang, H., & Braithwaite, J. (2017). Learning from incidents in health care: Critique from a safety-II perspective. *Safety Science, 99,* 115–121.

Sujan, M. A., Ingram, C., Mcconkey, T., Cross, S., & Cooke, M. W. (2011). Hassle in the dispensary: Pilot Study of a proactive risk monitoring tool for organisational learning based on narratives and staff perceptions. *BMJ Quality & Safety, 20,* 549–556.

Sujan, M. A., Koornneef, F., Chozos, N., Pozzi, S., & Kelly, T. (2013). Safety cases for medical devices and health it: Involving healthcare organisations in the assurance of safety. *Health Informatics Journal, 19,* 165–182.

Sutcliffe, K. M., Paine, L., & Pronovost, P. J. (2017). Re-examining high reliability: Actively organising for safety. *BMJ Quality & Safety, 26,* 248–251.

The Honourable Lord Cullen. (1990). Public inquiry into the piper Alpha disaster. London.

UK Ministry of Defence. (2007). *Defence Standard 00-56: Safety management requirements for defence systems*. London: The Stationary Office.

Vats, A., Vincent, C. A., Nagpal, K., Davies, R. W., Darzi, A., & Moorthy, K. (2010). Practical challenges of introducing who surgical checklist: UK pilot experience. *BMJ, 340.*

Vincent, C. A. (2004). Analysis of clinical incidents: A window on the system not a search for root causes. *Quality and Safety in Health Care, 13,* 242–243.

Wears, R. L. (2014). Risky business. *Annals of Emergency Medicine, 64,* 137–139.

Westbrook, J. I., Li, L., Lehnbom, E. C., Baysari, M. T., Braithwaite, J., Burke, R., Conn, C., & Day, R. O. (2015). What are incident reports telling us? A comparative study at two Austrian hospitals of medication errors identified at audit detected by staff and reported to incident system. *International Journal for Quality in Health Care, 21,* 1–9.

Chapter 3
Human Factors and Systems Approach to Patient Safety

Managing Human Error by Improving the Systems in Which People Work

3.1 Introduction

This chapter describes a human factors and systems approach to managing error in clinical systems. It is often suggested that 'human error' is the leading cause of adverse events. But what are the best strategies to understand and to manage error in clinical systems?

Historically, human error was thought to result from inexperience, from carelessness or from people simply not trying hard enough. This person-centred approach frequently led to blaming the individuals involved, but had limited success in improving patient safety. In the aftermath of major industrial accidents, human factors researchers developed a systems approach, which is gaining increasing popularity in healthcare (Reason 2000). This approach aims to identify and to manage deficiencies in clinical systems and organisational processes before such deficiencies create the conditions that set up people to fail.

The next section briefly introduces human factors as a discipline and outlines its application in healthcare. Then, we contrast the person-centred and the systems approaches to error. Subsequently, a practical human factors approach to managing error is described. A summary of the key lessons concludes the chapter.

3.2 Human Factors in Healthcare

Human factors is about understanding human characteristics, abilities and limitations in order to design and manage work in such a way that there is an optimal (or very good) fit between people and the work they do. Over the past decade, human factors have received a lot of interest from the healthcare community. However, this interest has often been limited to very specific aspects of human factors, such as physical ergonomics and behavioural safety approaches (Catchpole 2013).

P. Spurgeon et al., *Building Safer Healthcare Systems*,
https://doi.org/10.1007/978-3-030-18244-1_3

Physical ergonomics in the workplace is a part of human factors that many people will have come across. Physical ergonomics considers aspects such as the physical layout of workspaces, the physical characteristics of equipment and furniture, and the ergonomic properties of computer screens. An example of poor physical ergonomics might be a patient hoist where the wheels get stuck when used on carpet, prompting healthcare staff to abandon their use and handle patients without the hoist. This puts patients and staff at risk and should have been addressed in the design.

Behavioural safety approaches have been very popular in healthcare and are frequently based on team training and non-technical skills programmes developed in the aviation industry, where they go by the name of crew resource management (CRM) (Helmreich 2000). The roots of CRM were laid in 1979 in response to a NASA workshop on pilot error. While training of pilots originally had focused on the technical skills required to fly the aircraft, CRM introduced consideration of human factors and countermeasures to error in high-stress and high-risk environments. CRM is aimed at efficient team management, trying to enable teams to make good use of all available resources. Over the years, CRM has developed and evolved, and basic CRM training includes concepts such as team building, briefing strategies, situation awareness and stress management. In healthcare, one of the earliest and best-known implementations of CRM was in anaesthesia, where it was introduced as anaesthesia crisis resource management (Gaba et al. 2001). This included from the outset the use of simulation to create real-world training scenarios. Team training and non-technical skills training have since been adopted in a wide range of high-risk clinical domains, such as surgery, intensive care, and obstetrics, where they have made a positive contribution to reducing errors (Weaver et al. 2014).

However, there has been a tendency in healthcare to equate physical ergonomics and even more so team (and non-technical skills) training with human factors. This has been limiting and sometimes even counterproductive because such a perspective maintains the focus on the individual and their behaviour, while neglecting the wider systems aspects that impact on human performance (Russ et al. 2013). An increasingly popular conceptual model that attempts to overcome this limitation is the systems engineering initiative for patient safety (SEIPS) (Carayon et al. 2006). SEIPS integrates Donabedian's structural framework (Donabedian 1988) for examining the quality of care with a systems perspective of work. Donabedian's framework consisting of structure, process and outcome is well known in quality improvement circles, which might have increased the appeal of SEIPS. The basis of Donabedian's model is that the structure or the context within which care is delivered gives rise to specific care processes, which in turn produce particular outcomes. SEIPS uses this framework in a patient safety context to emphasise that structural issues such as the tools and equipment that people use, the design of the tasks and activities, the physical environment within which work is carried out, and the organisation and management of work (together forming the structure or context) give rise to care processes and patient safety and quality outcomes. In this way, SEIPS prompts us to consider not simply the individual and their behaviour (as done in behaviour safety approaches and team training), but rather the structure and the context of clinical work as a whole. Let us look in more detail at what a human factors and systems approach applied to error in clinical systems means.

3.3 Two Contrasting Views on Error in Clinical Systems

We know from the literature on preventable adverse events that errors occur regularly in clinical practice regardless of the setting (de Vries et al. 2008). We can assume that the true rate of error is much greater than studies looking only at adverse events suggest. For example, research investigating the reliability of different clinical processes found that many processes did not meet expected standards, with reliability rates frequently below 75% (Braithwaite et al. 2018; McGlynn et al. 2003). To clinicians who have been involved with clinical audit, such figures will hardly be surprising as audits often demonstrate significant differences between gold standards and actual practice.

There are different ways of approaching and managing this reliability shortfall and the significant number of errors in clinical systems: one is to manage the people who commit errors; the other, more fruitful approach is to manage the clinical systems and processes within which people work.

3.4 The Person-Centred Approach

When the delivery of care leads to patient harm, such as the death of a teenage girl following transplantation of organs of the wrong blood type (Hopkins Tanne 2003), it is important for organisations to understand what went wrong, and to provide reassurance to patients and their families that lessons have been learned (Sujan et al. 2017). If we look at the news coverage of adverse events, we frequently find that the explanation for what went wrong gives a single cause—human error. It is quite common to see headlines about, for example, doctors' errors, surgeons' mistakes and nurses' negligence that were allegedly what caused patient harm. A tragic case of wide significance in the NHS is the death of 6-year-old Jack Adcock, and the subsequent legal case against the doctor involved (see Box 3.1). The doctor was convicted of manslaughter and initially struck off the medical register (which she successfully appealed against), even though a wide range of failures in the system were identified.

Without doubt, the actions of doctors, surgeons, nurses and all the other healthcare professionals involved in providing care ultimately determine the quality of care and the safety of patients. But how successful is the focus on what individuals did wrong as a strategy for managing error, and how effective are the solutions that result from this approach in preventing future harm to patients?

The person-centred approach often is embedded in a blame culture, where people are held responsible uncritically for the errors they made. Individuals are singled out, disciplined, retrained and are reminded to pay more attention next time—if they are allowed to continue practising. This approach can be devastating for the members of staff involved (Wu 2000), and it prevents the organisation from looking beyond human error to other potential contributory factors (Sujan and Furniss 2015).

I once came across a case, where a nurse with more than twenty years of experience had been involved in a drug administration error that caused minor patient harm. In conversation, the nurse became very emotional, blaming herself for the error that she had committed. It was not just the nurse who blamed herself, however. As a consequence of the incident, the organisational response was to ban the nurse from administering medications until she had completed retraining and submitted a reflective diary. The nurse remarked that soon there would be no more nurses left on the ward who were allowed to give patients their medications.

This anecdotal experience will be all too familiar to many clinicians in different contexts. Research suggests that the blame culture significantly limits the effectiveness of organisational learning initiatives such as root cause analysis and incident reporting systems (Sujan 2015; Macrae 2015). Too often human error is regarded as a convenient end point for the analysis, and opportunities for deeper learning about organisational deficiencies are missed.

When one locates the cause of error with individual people, then the options for preventing and managing errors are limited (Sujan et al. 2016). It is not uncommon to see educational interventions and awareness-raising campaigns on the back of incident investigations, which are aimed at reminding staff of the importance of following certain protocols and procedures (Kellogg et al. 2017). Examples might include educating staff about the need for cognitive impairment assessment of elderly patients presenting to the emergency department, or reminding staff to complete a falls risk assessment. Such interventions are well intentioned, and they can be useful to a certain extent, but without accompanying system changes, any resultant improvement effects wear off quickly and are usually not sustainable.

Box 3.1. Death of Jack Adcock and the Bawa-Garba Legal Case

On 18 February 2011, 6-year-old Jack Adcock died of cardiac arrest as a result of sepsis at Leicester Royal Infirmary following a number of serious errors and omissions in his treatment. Dr. Hadiza Bawa-Garba, who was a junior paediatrician looking after Jack, was subsequently convicted of manslaughter on the grounds of gross negligence in 2015. Dr. Bawa-Garba was struck off the medical register in January 2018 following an intervention by the General Medical Council (GMC).The intervention by the GMC and the decision to strike Dr. Bawa-Garba off the medical register caused widespread concern among doctors. With this backing, Dr. Bawa-Garba successfully appealed against being struck off in August 2018.The death of Jack Adcock and the legal case against Dr. Bawa-Garba remain highly controversial. While it is recognised that

Dr. Bawa-Garba made mistakes, a large number of systems-based failures were identified, including communication problems, lack of adequate supervision, IT failures and local policies.

3.5 The Systems Perspective

Sidney Dekker explains the systems perspective on human error very intuitively and simply (Dekker 2014). He suggests that the analysis of incidents and adverse events should ask why it made sense to people at the time to act in the way they did. We can assume that most healthcare professionals are well trained and want to deliver good quality care to their patients. Why does a highly trained surgical team transplant organs of an incompatible blood type? Why does a nurse with more than twenty years of service administer the wrong dose of a drug? To offer human error as a cause runs the risk of missing and misrepresenting the complexity of everyday clinical work (Cook 2013). We need to look deeper, and try to understand and to explain why, at the time, the actions appeared to be perfectly reasonable to those involved.

Taking such a systems approach often reveals a second story of complexity, contradictions and necessary trade-offs (Cook 2013; Sujan et al. 2015). For example, multiple checks on organ compatibility might fail because responsibility for this task has not been allocated clearly, people are working under extreme time pressure, and assumptions about previous checks are made, which turn out to be invalid. Drug dose administration errors might result from staff shortages, which lead to staff having to take on additional duties and juggling several jobs concurrently, frequent interruptions and poor handovers due to time pressure.

Psychologist James Reason argued that while it is very hard to change the human condition, it is very feasible to change the conditions within which people work (Reason 2000). This line of reasoning forms the basis of the now famous Swiss cheese model of organisational accidents (Reason 1997). Reason explained that modern systems, such as nuclear power plants, commercial aircraft and also health systems are reasonably well protected against single failures through the use of barriers. If one barrier is penetrated, then the next one can stop the accident sequence. For example, in surgery there are multiple checks at different points in time and by different people to ensure that the correct surgery is performed on the correct surgical site on the right patient. If an earlier check fails, the error can be picked up during a subsequent check. Reason's argument is that for an accident to occur all of the barriers have to fail; i.e., accidents and adverse events usually result from multiple failures rather than from a single human error. There is, therefore, only rarely a single root cause to explain an accident, but rather there are several contributory factors all playing their part.

Reason likened barriers to slices of Swiss cheese, being full of holes. Holes in the barriers are weaknesses caused by unsafe acts by workers or through deficient organisational processes, which produce latent failure conditions. Reason turned his attention to those latent conditions and the deficient organisational processes that give rise to them. Examples of latent conditions include time pressure, distractions, unclear procedures and procedures that are out of date, broken computers and computer programmes that are difficult to use, and missing or inadequate equipment. These are the conditions that create hassle in everyday clinical work and that set up people to fail (Sujan 2012; Sujan et al. 2011).

The key contribution of Reason's model is the insight that deficient organisational processes produce latent failure conditions—and organisational processes can be managed; they are under an organisation's control. Typical deficiencies in organisational processes include, for example, low or inadequate staffing levels, poor planning of staffing rotas, inappropriate skills mix of staff on duty, inadequate maintenance schedules and routines, no processes to update and to maintain procedures and job descriptions, and procurement processes that fail to give attention to properly usability of technology. Once such deficiencies in organisational processes have been recognised, acknowledged and understood, they can be addressed and managed.

3.6 A Human Factors Approach to Managing Error

The systems perspective to understanding and managing errors is best implemented in practice through a human factors approach. There are numerous human factors techniques available, which can be used for modelling and representing tasks and processes, identifying and predicting vulnerabilities, and designing interventions (see, e.g., (Stanton et al. 2013), which describes about 100 different human factors methods). Each method has different strengths and weaknesses, and some are useful only for specific applications, while others can be used more generally. In this chapter, we focus on two specific approaches, which are aligned to expectations set out in the UK Health and Safety Executive (HSE) 'Human Factors Roadmap'. These approaches are hierarchical task analysis (HTA) as an example of a general-purpose task description method, and systematic human error reduction and prediction approach (SHERPA), which can be used to understand contributory factors in the work environment, which represent latent failure conditions.

3.7 Hierarchical Task Analysis

Hierarchical task analysis (HTA) was developed in 1967 as a method to represent the increasingly cognitive characteristics of human work activities, such as monitoring, anticipating and decision-making (Stanton 2006). HTA represents human activities based on a theory of goal-directed behaviour and includes a hierarchy of goals and subgoals linked by plans, which describe how subgoals combine to achieve the higher-level goal. Plans can be used to express any kind of algorithm, e.g. simple

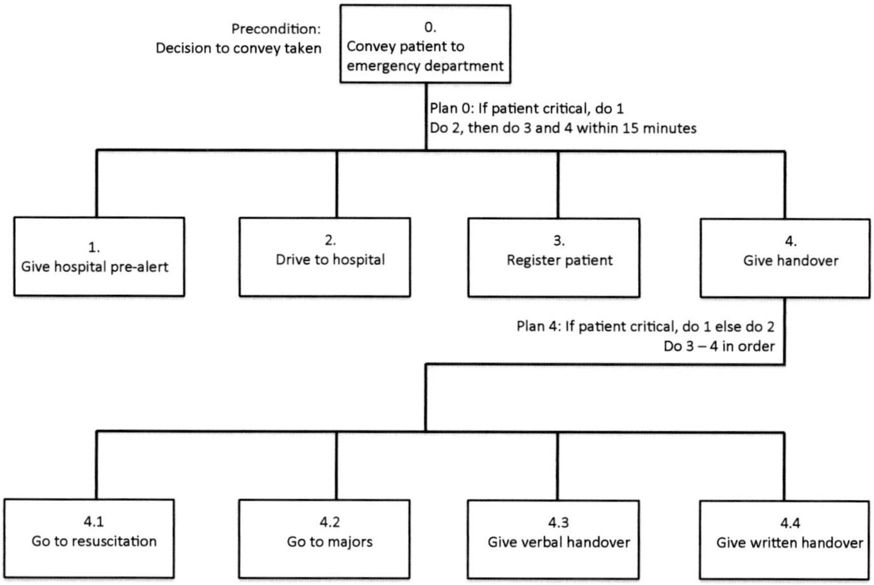

Fig. 3.1 Example of HTA analysis

sequential ordering (such as do Step 1 to Step 3 in order), free ordering (do Steps 1, 2, 3 in any order), as well as more complex loops (such as do Step 1 and Step 2 in order until signal A is active, then do Step 3). This representation creates a tree-like structure, where the leaves represent task steps that are considered elementary (e.g. basic manual operations) or where further decomposition is not considered necessary.

A simple example of the use of HTA to represent the task of an ambulance crew conveying a patient to the emergency department (ED) is given in Fig. 3.1. The high-level goal is to convey the patient to the ED once the decision to convey has been taken (precondition). This goal can be broken down into a number of lower-level goals, namely to give a pre-alert if the patient is critical, and then to drive to the ED, and to register and to hand over the patient within the 15-min timeframe set within the NHS in England. All of these tasks can be broken down further depending on the needs of the analysis. In Fig. 3.1, only the handover task is broken down further. In order to hand over the patient, the ambulance crew has to either go to resuscitation (critical patient) or to the 'major' area, and then provide a verbal handover and leave the written handover form with ED staff.

HTA has proven to be a very versatile human factors method, which has been applied in many different contexts over the past fifty years (Stanton 2006), including healthcare (Chana et al. 2017; Lane et al. 2006; Parand et al. 2017). The main strengths of HTA are the flexible hierarchical decomposition, which allows activities to be broken down to the level that is considered adequate for the purpose of the analysis, and the explicit representation of algorithmic plans.

Another strength of HTA, which is particularly relevant in a healthcare context, is its ability to support clinical teams in defining and understanding clinical processes, which hitherto had not been formally designed or documented. The use of HTA can also provide team members from different backgrounds with the opportunity to build important relationships with each other, which in normal clinical practice they would not have. Creating opportunities to strengthen the social infrastructure of safety and enhancing staff engagement should be a key patient safety improvement strategy (Sujan 2015). The key features of HTA are described in Chap. 5 within the overall SCS framework, and a fully worked example of its use is given in the SCS implementation chapter (Chap. 6).

3.8 Systematic Human Error Reduction and Prediction Approach

The systematic human error reduction and prediction approach (SHERPA) was originally developed to analyse and reduce errors in the nuclear and process industries, but has been used since in many other contexts (Embrey 1986). It is similar in structure to failure mode and effects analysis (FMEA) (DeRosier et al. 2002), but it is based on a simple taxonomy of human errors, which can function as a guide for the identification of failure modes.

SHERPA uses the HTA representation and systematically analyses the basic task steps, i.e. the bottom leaves in the HTA tree diagram. The analyst classifies each basic task step according to the behaviour type and then applies the corresponding human error modes. The suggested behaviour types are action, checking, information retrieval, communication and selection. Basic human error modes for each of these behaviour types are shown in Table 3.1.

Table 3.2 provides an example of the failure analysis of one of the bottom leaves: task 'give verbal handover' (task step 4.3). The guidewords for communication (I01–I04) are applied to generate credible error modes. Each error mode is then rated for its likelihood and the severity of the potential consequences. Different rating schemes are available, but a commonly used one (also used in the example) is the 5 × 5 matrix, where likelihood and severity are rated on a scale from 1 (rare occurrence; negligible consequences) to 5 (frequent occurrence; patient death). The risk score is obtained by multiplying the scores for likelihood and severity. In this way, the error modes can be prioritised for investigation and action based on their risk score.

The SHERPA human error taxonomy provides a simple grouping of human errors that are likely to be influenced by similar contextual conditions. These contextual conditions are referred to as performance influencing factors (PIF) or performance

Table 3.1 SHERPA human error taxonomy

Behaviour type	Code	Error mode
Action	A01	Action too long/too short
	A02	Action mistimed
	A03	Action in wrong direction
	A04	Action too little/too much
	A05	Action too fast/too slow
	A06	Misalign
	A07	Right action on wrong object
	A08	Wrong action on right object
	A09	Action omitted
Checking	C01	Check omitted
	C02	Check incomplete
	C03	Right check on wrong object
	C04	Wrong check on right object
	C05	Check too early/too late
Information retrieval	R01	Information not obtained
	R02	Wrong information obtained
Communication	I01	Information not communicated
	I02	Wrong information communicated
	I03	Information communication incomplete
	I04	Information communication unclear
Selection	S01	Selection omitted
	S02	Wrong selection

shaping factors (PSF). SHERPA promotes a systems-based approach to understanding human error through consideration of PIFs. For example, task step 4.3 'give verbal handover' is a communication task. Communication tasks are likely to be influenced by the condition of the work environment (e.g. noise levels, distractions), the availability of structured communication protocols and the extent to which the communication partners have shared information needs. The final step of the SHERPA analysis is to assess the status of the PIFs for a specific situation and to propose improvements based on this. For example, if there are frequent interruptions to the

Table 3.2 Error analysis example

Task step	Error mode	Likelihood	Consequences	Risk score
4.3 Give verbal handover	I01—No verbal handover takes place	2	4	8
	I02—Wrong mechanism of injury communicated	2	4	8
	I03—No information about drug allergies available	3	4	12
	I04—N/A	–	–	–

handover, the use of a dedicated handover location that is sheltered from excessive noise and interruptions might be considered.

SHERPA does not propose a set of PIFs to consider, but in the human factors literature, there are several types of lists of PIFs, which are in common use. These vary in the degree of their specificity to particular tasks and forms of behaviour. For example, the set of PIFs proposed in the human error assessment and reduction technique (HEART) comprises around 40 very specific factors (called error-producing conditions) (Williams 1988). The cognitive reliability and error analysis method (CREAM) on the other hand only suggests consideration of 9 more abstract PIFs, called common performance conditions (Hollnagel 1998). Arguably, the best practical guidance is to rely on frontline experience as a source of evidence, and to use PIF lists as prompts at the level that appears most appropriate to the domain and task under analysis.

The failure analysis is usually conducted within multidisciplinary groups that include all relevant stakeholders. SHERPA can facilitate and structure the group work, and can support participants to think differently about what they do, and the vulnerabilities that they face (Sujan et al. 2018). This can help to generate systems-based rather than person-focused improvement suggestions.

3.9 Summary

This chapter contrasted the person-centred and the systems-based approach to understanding and managing error in clinical systems. The person-centred approach focuses on the individual, often suggests improvements based on education and training and tends to impose disciplinary action on staff involved in incidents. The person-centred approach often operates, therefore, within a blame culture. On the other hand, the systems-based approach aims to understand and to improve the systems within which people work without necessarily attributing blame to individuals.

We have described two common human factors methods for representing and assessing human activities in the context of error analysis: hierarchical task analysis and systematic human error reduction and predication approach. The strength of these methods is that they facilitate the assessment of work by multidisciplinary groups, thus encouraging engagement with and ownership of the analysis. They also draw attention to the role of performance influencing factors, which can be the starting point for improvement suggestions.

There are more than 100 human factors methods available, depending on the aim of the analysis, the domain under consideration and the level of expertise of the analyst. This can be confusing and daunting. The two methods described in this chapter are used frequently and require only a moderate amount of training. They are, therefore, well suited for use within clinical environments.

Criticisms have been raised about the lack of evidence that such methods lead to improvements that are superior in practice than those generated from other approaches, such as Lean and quality improvement techniques (Dean Franklin et al. 2012; Shebl et al. 2009). Such evidence has proven to be hard to deliver because the analysis of safety problems is only the starting point for improving patient safety. Much depends on how improvements are taken forward, implemented and sustained. This usually requires a strong safety culture and an organisation that is committed to improving patient safety.

References

Braithwaite, J., Hibbert, P. D., Jaffe, A., et al. (2018). Quality of health care for children in Australia, 2012–2013. *Jama, 319,* 1113–1124.

Carayon, P., Hundt, A. S., Karsh, B., Gurses, A. P., Alvarado, C., Smith, M., et al. (2006). Work system design for patient safety: The Seips model. *Bmj Quality & Safety, 15,* I50–I58.

Catchpole, K. (2013). Spreading human factors expertise in healthcare: Untangling the knots in people and systems. *Bmj Quality & Safety, 22,* 793–797.

Chana, N., Porat, T., Whittlesea, C., & Delaney, B. (2017). Improving specialist drug prescribing in primary care using task and error analysis: An observational study. *British Journal of General Practice, 67,* E157–E167.

Cook, R. (2013). Resilience, the second story, and progress on patient safety. In E. Hollnagel, J. Braithwaite, & R. Wears (Eds.), *Resilient health care.* Farnham: Ashgate.

de Vries, E. N., Ramrattan, M. A., Smorenburg, S. M., Gouma, D. J., & Boermeester, M. A. (2008). The incidence and nature of in-hospital adverse events: A systematic review. *Quality and Safety in Health Care, 17,* 216–223.

Dean Franklin, B., Shebl, N. A., & Barber, N. (2012). Failure mode and effects analysis: Too little for too much? *Bmj Quality & Safety, 21,* 607–611.

Dekker, S. (2014). *The field guide to understanding human error.* Farnham: Ashgate Publishing Ltd.

Derosier, J., Stalhandske, E., Bagian, J. P., & Nudell, T. (2002). Using health care failure mode and effect analysis: The Va National Center for patient safety's prospective risk analysis system. *Joint Commission Journal on Quality Improvement, 28*(248–67), 209.

Donabedian, A. (1988). The quality of care: How can it be assessed? *Jama, 260,* 1743–1748.

Embrey, D. (1986). Sherpa: A systematic human error reduction and prediction approach. In *Proceedings of the International Topical Meeting on Advances in Human Factors in Nuclear Power Systems*. Knoxville, Tennessee: American Nuclear Society.

Gaba, D. M., Howard, S. K., Fish, K. J., Smith, B. E., & Sowb, Y. A. (2001). Simulation-based training in anesthesia crisis resource management (Acrm): A decade of experience. *Simulation & Gaming, 32,* 175–193.

Helmreich, R. L. (2000). On error management: Lessons from aviation. *BMJ, 320,* 781–785.

Hollnagel, E. (1998). *Cognitive reliability and error analysis method (cream).* Oxford: Elsevier.

Hopkins Tanne, J. (2003). When Jesica died. *Bmj: British Medical Journal, 326,* 717.

Kellogg, K. M., Hettinger, Z., Shah, M., Wears, R. L., Sellers, C. R., Squires, M., et al. (2017). Our current approach to root cause analysis: Is it contributing to our failure to improve patient safety? *Bmj Quality &Amp Safety, 26,* 381–387.

Lane, R., Stanton, N. A., & Harrison, D. (2006). Applying hierarchical task analysis to medication administration errors. *Appl Ergon, 37,* 669–679.

Macrae, C. (2015). The problem with incident reporting. *Bmj Quality & Safety.*

Mcglynn, E. A., Asch, S. M., Adams, J., Keesey, J., Hicks, J., Decristofaro, A., et al. (2003). The quality of health care delivered to adults in the United States. *New England Journal of Medicine, 348,* 2635–2645.

Parand, A., Faiella, G., Franklin, B. D., Johnston, M., Clemente, F., Stanton, et al. (2017). A prospective risk assessment of informal careers' medication administration errors within the domiciliary setting. *Ergonomics*, 1–18.

Reason, J. (1997). *Managing the risks of organizational accidents.* Farnham: Ashgate.

Reason, J. (2000). Human error: Models and management. *Bmj, 320,* 768–770.

Russ, A. L., Fairbanks, R. J., Karsh, B.-T., Militello, L. G., Saleem, J. J., & Wears, R. L. (2013). The science of human factors: Separating fact from fiction. *Bmj Quality & Safety, 22,* 802–808.

Shebl, N. A., Franklin, B. D., & Barber, N. (2009). Is failure mode and effect analysis reliable? *Journal of Patient Safety, 5.*

Stanton, N. (2006). Hierarchical task analysis: Developments, applications, and extensions. *Applied Ergonomics, 37.*

Stanton, N., Salmon, P. M., & Rafferty, L. A. (2013). *Human factors methods: A Practical guide for engineering and design.* Ashgate Publishing, Ltd.

Sujan, M. (2015). An organisation without a memory: A qualitative study of hospital staff perceptions on reporting and organisational learning for patient safety. *Reliability Engineering & System Safety, 144,* 45–52.

Sujan, M., & Furniss, D. (2015). Organisational reporting and learning systems: Innovating inside and outside of the box. *Clinical Risk, 21,* 7–12.

Sujan, M., Pozzi, S., & Valbonesi, C. (2016). Reporting and learning: From extraordinary to ordinary. In J. Braithwaite, R. Wears, & E. Hollnagel (Eds.), *Resilient health care: Reconciling work-as-imagined with work-as-done* (Vol. 3). Farnham: Ashgate.

Sujan, M., Spurgeon, P., & Cooke, M. (2015). The role of dynamic trade-offs in creating safety—A qualitative study of handover across care boundaries in emergency care. *Reliability Engineering & System Safety, 141,* 54–62.

Sujan, M. A. (2012). A novel tool for organisational learning and its impact on safety culture in a hospital dispensary. *Reliability Engineering & System Safety, 101,* 21–34.

Sujan, M. A., Embrey, D., & Huang, H. (2018). On the application of human reliability analysis in healthcare: Opportunities and challenges. *Reliability Engineering & System Safety.*

Sujan, M. A., Huang, H., & Braithwaite, J. (2017). Learning from incidents in health care: Critique from a Safety-II perspective. *Safety Science, 99,* 115–121.

Sujan, M. A., Ingram, C., Mcconkey, T., Cross, S., & Cooke, M. W. (2011). Hassle in the dispensary: Pilot study of a proactive risk monitoring tool for organisational learning based on narratives and staff perceptions. *Bmj Quality & Safety, 20,* 549–556.

Weaver, S. J., Dy, S. M., & Rosen, M. A. (2014). Team-training in healthcare: A narrative synthesis of the literature. *Bmj Quality & Safety, 23,* 359–372.

Williams, J. C. (1988). A data-based method for assessing and reducing human error to improve operational performance. In *Conference Record for 1988 IEEE Fourth Conference on Human Factors and Power Plants*, June 5–9, 1988, 436–450.

Wu, A. W. (2000). Medical error: The second victim. *Bmj, 320,* 726–727.

Chapter 4
Safety and Culture: Theory and Concept

4.1 Introduction

The previous chapters have looked at the thinking behind the Safer Clinical Systems approach, and current concerns about patient safety and about what the health sector might learn from other sectors and from human factors in particular. Perhaps, implicit in much writing about patient safety is the notion of organisational culture, or safety culture and whether this is seen as negative and culpable for failure (as in Mid-Staffordshire NHS Foundation Trust and documented in the Francis Enquiry Report 2010), or as advocated in a positive sense as the underpinning to improved patient safety. It has to be said that many of the proponents of safety culture as a key element of improved patient safety rarely provide any great precision over just what this cultural framework looks like, how it is created and just how it flows through to behaviour that changes patient outcomes. There is often a degree of tautology in the way safety culture is discussed and this can leave the practitioner seeking to implement change somewhat confused. Do specific safety initiatives, for example, to reduce medication errors create a sense of heightened awareness about safety and hence promote an improved culture around safety? Or does a drive to demonstrate the value and importance of a robust safety culture create a receptivity that enables specific initiatives to succeed. The directionality of how improved cultures develop may seem an academic nicety but clarification could have important practical implications for initiating safety improvements. The safety culture work within the SCS programme attempted to tackle many of these missing aspects. However, before looking at this practical application of safety culture (Sect. 2, Chap. 7), this chapter will examine some of the concepts involved in organisational and safety culture, and the evidence relating safety culture to improvements in patient safety.

P. Spurgeon et al., *Building Safer Healthcare Systems*,
https://doi.org/10.1007/978-3-030-18244-1_4

4.2 What Is Understood by the Term Safety Culture?

In contrast to organisational culture, the term safety culture is relatively new. Halligan and Zecevic (2010) attribute its emergence to shortly after the Chernobyl nuclear power disaster in 1988. It began to be used in the healthcare sector from 2000 onwards although rather loosely in terms of precise meaning or definition. Halligan and Zecevic (op cit) recognising this imprecision undertook a systematic review of the term and its use in healthcare. A total of 139 studies were included on the basis of providing sufficient detail or definition of the meaning of safety culture. They report that many studies use the terms safety culture and safety climate interchangeably despite others arguing that they are conceptually distinct. Perhaps, the most pragmatic approach is represented by Gaba et al. (2003) and Colla et al. (2005) who suggest that safety climate can be thought of as the surface feature of culture and more accessible in terms of measurement. However, one can see why the confusion remains as they identify aspects such as attitudes and perceptions, but the same terms are often used to outline the content of safety culture. The number of studies failing to provide any definitions far outweighed those that did. This contributes to the slightly disturbing sense of 'we all know what we mean' but actually we do not. Such vagueness is unlikely to help practitioners in their attempts to change culture and this may well explain much of the relative failure of initiatives in this area.

Mannion and Davies (2013) pick up on this and raise the consequential question of whether prescriptions or demands to change culture will have any impact. This point was made in the wake of the failures at Mid-Staffordshire NHS Foundation Trust and the resulting Francis Report (2013). Dixon-Woods et al. (2013) sought to examine the problem of advocated but ill specified culture change on a national scale given that the Francis Report emphasised the negative culture at Mid-Staffordshire citing short-term decisions and policies, poor leadership, tolerance of poor standards. Dixon-Woods and colleagues wonder how many other organisations may be experiencing similarly worrying cultural conditions. From a very large multi-method study they concluded that NHS organisations were hampered in developing positive supportive environments by:

(a) a lack of clear organisational goals that might serve to focus attention on achievements and encourage greater cohesion in the workforce
(b) a sense of organisations being overwhelmed by the need to meet the requirements of competing for regulatory bodies
(c) poorly designed and often a variably functioning system that failed to properly support staff in their efforts to deliver good quality care.

It is understandable why cultures fostered by such inadequate provisions can create performance levels that fall far short of that needed in the context of high-quality and safe healthcare delivery. In the sense of a mirror image, it is though possible to see some encouragement in the notion of how a really positive and supportive culture might sustain improvements, and indeed be necessary in order to do so.

Simon and Cistaro (2009, p. 30) assert that 'safety excellence is a product not only of the right programmes but also the right culture'. In a similar, vein Krause and Hidley (2009, p. 132) make a powerful statement about the responsibility of those in leadership roles in organisations. They say that in a positive and proactive safety culture individual employees should not be accountable 'without giving them the resources, information, leadership support and encouragement they need for success'. In the USA, Hudson et al. (2009) report that The Joint Commission (2009) has included in an update of its leadership standards the requirement for leaders to 'create and maintain a culture of safety and quality throughout the hospital' (Joint Commission 2009 Leadership standard LD 3.10). Despite the claims that culture lacks clear definition there seems to be some consensus at a fairly broad level that culture reflects a combination of some mental constructs (attitudes, values, beliefs), a set of norms (behaviour, practices), institutional structures (committees, arrangements) and artefacts (processes, equipment), (Hudson et al. 2009, p. 3). How each of these areas manifests or impacts upon shaping culture in a particular organisation is less clear. The interesting dynamic is how an existing culture drives the behaviour of staff in relation to promoting or inhibiting patient safety. A strong, powerfully positively directed safety culture may well be key to promoting effective safety conscious behaviour in staff—indeed this is the thrust of the argument in this chapter. But equally, it is important to acknowledge that a culture, albeit strong, that emphasises say financial performance above safety may contribute to a lack of focus on appropriate safety behaviours.

Singla et al. (2006) provide an operational approach to differentiating safety culture from safety climate-seeing the former as based upon deep-seated and underlying values, while climate and especially climate surveys represent a current snapshot of how these cultural forces are manifesting at a particular period of time. It may be that in healthcare organisations such as the NHS have a very powerful and well-established professional (especially medical) culture and therefore attempts to change culture often founder on this or result in more superficial climate changes but leave the dominant culture unchanged (Broadbent et al. 1992).

4.3 Safety Culture and Links to Organisational Performance

The issue of incident reporting has become entwined with the notion of patient safety culture particularly as systems and procedures for reporting have become established. Reporting rates are in themselves a problem since staff feel uncomfortable in many circumstances. Several studies have reported that a positive safety culture can help to overcome this reluctance (Waring 2005; Evans et al. 2006; Benn et al. 2009). Burlison et al. (2016) have recently sought to differentiate particular aspects of patient safety culture that relate to incident-reporting rates. The components from their analysis that best predicted reporting were—feedback about errors, management support for patient safety, organisational learning, non-punitive response to error and teamwork within units. This was a large-scale analysis and is consistent with the previously mentioned studies. Hudson et al. (op cit) classified aspects of patient safety culture into three broad areas—Patient Safety Norms and Behaviours, Teamwork and Communication and Job Attitudes. Using this broad model, they report a number of important findings of the operation of patient safety culture, notably that it can influence behaviour at various levels within a hospital from a small group, specific units and also hospital-wide. They also quote other significant findings such as differential ratings of teamwork which in itself can influence patient safety behaviour (Sexton et al. 2006), and that patient safety culture in two ICU units was associated with sustained improvements in medication errors, length of stay and nursing turnover rates (Pronovost et al. 2003).

A further strand of thinking relates to how patient safety culture may relate to more generic quality improvement initiatives. A number of studies have reported that they are different and may be contrastingly affected by generic approaches. Indeed McFadden, Stock and Gowen (2014) found a negative impact of a continuous quality improvement (CQI) initiative and hospital-acquired conditions (HACs). This is similar to Douglas and Fredenhall (2004) who suggest that CQI initiatives are concerned with processes, including efficiencies and standardisation, and that more specific safety outcomes may not always be positively affected by these approaches. However, Kristensen et al. (2015) did find a positive impact of implementing quality management systems on safety climate. It is, perhaps, unsurprising that leadership in organisations is reported to play a major part in improving patient safety culture. McFadden, Stock and Gowen (op cit) report this as do Kristensen et al. (2016). They emphasise particularly the inspirational leadership implicit in transformational leadership models.

Yu et al. (2016) suggest that culture has been under-researched in terms of its potential contribution to safe practice. Perhaps, one of the most useful accounts of the role of culture is provided by the Institute for Healthcare Improvement (Frankel et al. 2017) who suggest that culture is made up of some key components:

(a) Psychological safety (the sense of being able to challenge or ask questions without repercussions
(b) Accountability (being appropriately equipped and expected to act in a safe manner)
(c) Teamwork and Communication (a shared view about how we go about delivering care and responding to issues)
(d) Negotiation (seeking authentic communication and agreement about important issues).

Each of these facets they argue must be nurtured and developed by effective leadership- thus importantly giving impetus to the notion that there needs to be a focus in every organisation of someone (a leader but not necessarily the Chief Executive) who focuses minds and brings staff together around patient safety.

The overall question of whether and how patient safety culture is associated with improved patient outcome remains contentious. Some recent papers demonstrate this in arguing strongly in opposing directions. Braithwaite et al. (2016) present a research protocol which they believe is necessary to resolve the issue. Although this article acknowledges the potential and asserted benefits of a positive patient safety culture they cite a review by Parmelli et al. (2011) as not finding the case proven and therefore conclude themselves that 'Despite these potential consequences we do not know with sufficient confidence about the association between organisational and workplace cultures, and patient outcomes, in healthcare environments' (p. 1).

Contrasting conclusions are to be found in a review by Sacks et al. (2015) relating to surgical procedures. They report improvements in patient safety culture linked to a range of outcomes from post-operative complications, reduction in surgical morbidity as well as increased use of appropriate prophylactic measures.

Vogus (2016) in a more discursive paper, perhaps, nicely sketches out the present position. He argues strongly that we need to see safety culture as much more differentiated between different units within hospital settings, and that failure to recognise this may account for some of the inconsistency in findings. Moreover, he also notes that different facets of safety culture may be more or less influential in these different settings. The proposition that patient safety culture needs to be examined in a more nuanced way taking into account differences within the organisation and particularly the strength and consistency of cultural values accords with the approach taken within the Safer Clinical Systems project. The particular approach and measures used in the Safer Clinical System project are described in Sect. 2, Chap. 7 in terms of the development and use of the Safety Culture Index (SCI).

References

Benn, J., Koutantji, M., Wallace, L., et al. (2009). Feedback from incident reporting: Information and action to improve safety. *Quality and Safety in Health Care, 18,* 11–21.

Braithwaite, J., Herkes, J., Ludlow, K., Lamprell, G., & Testa, L. (2016). Association between organisational and workforce cultures, patient outcomes: Systematic review protocol. *BMJ Open*. https://doi.org/10.1136/bmjopen-2016-013758.

Broadbent, J. R., Laughlin, R., & Shearn, D. (1992). Recent financial and administrative changes in general practice: An unhealthy intrusion into medical autonomy? *Financial Accountability & Management, 8*(2), 129–148.

Burlison, J. D., Quillivan, R. R., Kath, L. M., Zhou, Y., Courtney, S. C., Cheng, C., et al. (2016). A multilevel analysis of U.S. hospital patient safety culture relationships with perceptions of voluntary event reporting. *Journal of Patient Safety*, 1–7.

Colla, J. B., Bracken, A. C., Kinney, L. M., et al. (2005). Measuring patient safety climate: A review of surveys. *Quality and Safety in Health Care, 14,* 364–366.

Dixon-Woods, M., Baker, R., Charles, K, Dawson, J., Jerzembek, G., Martin, G., et al. (2013). Culture & behaviour in the English National Health Service; overview of lessons from a large multi-method study. *BMJ Quality & Safety*. https://doi.org/10.1136/bmjqs.2013.001947.

Douglas, T. J., & Fredendall, L. D. (2004). Evaluating the Denning management model of total quality in services. *Decision Sciences, 35*(3), 393–422.

Evans, S. M., Berry, J. G., Smith, B. G., et al. (2006). Attitudes and barriers to incident reporting: A collaborative hospital study. *Quality and Safety in Health Care, 15,* 39–43.

Francis, R. (2013). *Report of the mid-staffordshire NHS Foundation Trust public inquiry*. London: Stationery Office.

Frankel, A., Haraden, C., Federico, F., & Henoci-Edwards, J. (2017). A framework for safe, reliable and effective care. *White paper*. Cambridge, MA: Institute for Healthcare Improvement and Safe and Reliable Healthcare.

Gaba, D. M., Singer, S. J., Sinako, A. D., et al. (2003). Differences in safety climate between hospital personnel & naval aviations. *Human Factors, 45,* 173–185.

Halligan, M., & Zecevic, A. (2010). Safety culture in healthcare: A review of concepts, dimensions, measures and progress. *BMJ Quality & Safety*. https://doi.org/10.1136/bmjqs.2010.040964.

Hudson, D. W., Sexton, J. B., Thomas, E. J., & Berenholtz, S. M. (2009). A safety culture primer for the critical care clinician. The Role of Culture in Patient Safety & Quality Improvement. *Contemporary Critical Care, 7*(5), 1–14.

Krause, T., & Hidley, J. (2009). *Taking the lead in patient safety: How healthcare leaders influence behaviour and create culture*. Hoboken, N.Y: Wiley.

Kristensen, S., Hammer, A., Bartels, P., Sunol, R., Groene, O., Thompson, C. A., et al. (2015). Quality management & perceptions of teamwork and safety climate in European hospitals'. *International Journal for Quality in Health Care, 27*(6), 499–506.

Kristensen, S., Christensen, K. B., Jaquet, A., Beck, C. M., Sabroe, S., Bartels, P., et al. (2016). Strengthening leadership as a catalyst for enhanced patient safety culture: A repeated cross-sectional experimental study. *BMJ Open, 6*. https://doi.org/10.1136/bmjopen-2015-010180.

Mannion, R., & Davies, H. (2013). Will prescriptions for cultural change improve the NHS? *British Medical Journal, 346,* 1305.

McFadden, K. L., Stock, G. N., & Gowen, C. R. III. (2014). Leadership, safety climate, & continuous quality improvement. Impact on process quality and patient safety. *Health Care Management Review, 40*(1), 1–11.

Parmelli, E., Flodgren, G., Beyor, F., et al. (2011). The effectiveness of strategies to change organisational culture to improve healthcare performance: A systematic review. *Implement Sci, 6,* 1–8.

Pronovost, P. J., Weast, B., Holzmueller, C. G., et al. (2003). Evaluation of the culture of safety: Survey of clinicians & managers in an academic medical centre. *Quality and Safety in Health Care, 12*(6), 405–410.

Sacks, G. D., Shannon, E. M., Dawes, A. J., Rollo, J. C., Nguyen, D. K., Russell, M. M., et al. (2015). Teamwork, communication and safety climate: A systematic review of interventions to improve surgical culture. *BMJ Quality & Safety*. http://dx.doi.org/10.1136/bmjqs-2014-003764.

Sexton, J. B., Thomas, E. J., & Helmreich, R. L. (2006). Error, stress & teamwork in medicine & aviation: A cross-sectional study. *Chirurg, 71*(6), 138–142.

Simon, S. I., & Cistaro, P. A. (2009). Transforming safety culture: Grassroots-led/management—Supported change at a major utility. *Professional Safety, 54*(4), 28–35.

Singla, A. K., Kitch, B. T., Weissman, J. S., & Campbell, E. G. (2006). Assessing patient safety culture: A review and synthesis is of measurement tools. *Journal of Patient Safety, 2,* 105–115.

The Joint Commission. (2009). *Comprehensive Accreditation Manual for Hospitals (AMH). The Official Handbook*. Chicago: Joint Commission Resources (2008).

Vogus, T. (2016). Safety climate strength: A promising construct for safety research and practice. *BMJ Quality & Safety*. https://doi.org/10.1136/bmjqs-2015-004847.

Waring, J. (2005). Beyond blame: Cultural barriers to medical incident reporting. *Social Science and Medicine, 60,* 1927–1935.

Yu, A., Flott, K., Chainani, N., Fontana, G., & Darzi, A. (2016). *Patient safety 2030*. London, UK: NIHR Imperial Patient Safety Translational Research Centre.

Chapter 5
An Outline of the Evolution and Conduct of the Safer Clinical Systems Programme

Background to the Approach

As noted in earlier chapters, the one key thing that other safety-critical industries do that healthcare does not generally do, is that they search proactively to identify hazards and assess their risk. As a result, they pre-emptively control or manage the risk in order to obtain safe outcomes. The Safer Clinical Systems approach embodies this philosophy. There are some crucial lessons here that can inform the paradigm shift needed for the NHS to become proactive in building safety systems for our patients.

In this chapter, we describe in outline the development of our approach, offer a structured step-by-step method for creating safer systems for our patients, describe the purpose of each step, what you might expect to gain from each of them, and briefly identify the tools and techniques you could use to support each step. More detailed information can be found in 'Safer Clinical Systems, A new, proactive approach to building safe healthcare systems. A reference guide for clinicians and managers (Cooke et al. 2016).' An account of the implementation of the Safer Clinical System model is given in Part II, Chap. 6.

5.1 The Development of the Approach

Background of the Ideas

The idea of taking a new look at how the NHS manages the safety of patients was conceived by The Health Foundation who put out a tender document seeking organisations interested in developing the concept. They wanted a core team of experts in safety to work with selected NHS Trusts to develop a different approach to safety drawing on the learning from High-Reliability Organisations [HROs]. This was put out to competitive tender in 2008. The team that won the tender was led by Warwick Medical School and included NHS staff with expertise in patient safety, service and system improvement and people from outside the NHS who had experience of work-

P. Spurgeon et al., *Building Safer Healthcare Systems*,
https://doi.org/10.1007/978-3-030-18244-1_5

ing with HROs, Human Factors and technical expertise in using a range of tools and techniques associated with safety improvement.

The Safer Clinical Systems programme was developed by the Warwick team to explore a new approach to safety improvement programmes. Unlike previous programmes, which had been developed around specific safety issues and involved closely defined interventions in confined settings, the Safer Clinical Systems programme combined methods from improvement science and safety engineering *to identify risk* ***within a system*** using a formalised, prospective approach to safety, to assess these risks and then develop interventions along the patient pathway linked to other factors influencing performance in that system.

The traditional approach of the NHS has been focussed on reacting to harm. The NHS needs an approach that is continuously aiming not only to reduce harm but to do this by searching out and minimising risk, and to develop resilience to unexpected challenges. To achieve this, a 'systems approach' to patient safety is required. A safer clinical system is a clinical system that 'delivers high-quality care to the patient, is demonstrably free from unacceptable levels of risk and has the resilience to withstand known and unexpected variations and challenges'.

Safer Clinical Systems, at its inception, was a new approach to improving patient safety in the context of healthcare based on engendering a learning ethos and promoting engagement through utilising learning from high reliability organisations, i.e. one which is continuously aiming not only to improve process reliability and to reduce harm, but also to search out and minimise risk, therefore developing resilience.

Retrospectively, the programme can be seen as embodying two phases

Phase 1—Proof of concept
Phase 2—Application and testing of the concept.

5.2 Summary of Phase 1 September 2008 to December 2010

Phase 1 of the Safer Clinical Systems programme targeted healthcare systems recognised by the participating Trusts to be vulnerable to the occurrence of safety issues. These were assessed using a range of tools, to identify key areas of a safety risk, assess reliability and design focused interventions to reduce the level of risk by improving process reliability. The Trusts were also selected through a tendering process and 'awarded' a sum of money to undertake the necessary work as part of the overall programme [Sometimes referred to as Award Holders].

There was an aspect of action research for the programme. Award Holders and the Warwick team took an iterative approach to allow exploration of diverse issues in the context of both the system and the organisational environment, with an assessment of the impact of the project focussing on levels of reliability. This also allowed tools and approaches to be developed, tested, debated and modified to help formulate Phase 2.

Phase 1 demonstrated proof-of-concept for the approach and showed preliminary evidence of the ability to improve reliability in four systems under study:

Aspects of patient flow within the hospital environment
Patient information transfer
Patient handover between different care settings
Medication prescribing within the hospital environment.

Phase 1 generated a considerable body of learning applicable to the broader Safer Clinical Systems programme which was used to guide the development of Phase 2, particularly in terms of refining methodological and measurement processes and the wider goal of engendering a sustainable safety culture at the team and organisational level.

Summary of key learning from Phase 1

- Adopt a systematic approach
- Align improvement at all levels
- Culture and relationship changes required as well as technical
- Link between unreliable systems and safety
- Engage, motivate and involve staff
- Work across organizational boundaries
- Leadership and direction is vital

http://www.health.org.uk/areas-of-work/programmes/safer-clinical-systems/learning/

During Phase 1 of Safer Clinical Systems, it became clear that redesigning a system is not facilitated by dividing it into isolated parts; the whole also needs to be considered by all participants, which we describe as the 'zoom in—zoom out' nature of the Safer Clinical Systems approach. A system can be defined as 'a clinical pathway of care and the factors that influence that pathway, both within and without the organisation'. We 'zoom in' to fix part of the microsystem which may then require an intervention in the wider system, the context for the microsystem, so we need to 'zoom out'.

In addition, work on human factors and organisational culture has shown that Safer Clinical Systems projects have the capacity to influence culture by a process of spread that can modify the working practices of the entire organisation. The Award Holder programme(s) for the Trust teams set out to promote horizontal learning between teams and programmes and vertical learning and change throughout the hierarchy of an organisation.

System's thinking is, therefore, one of the defining features of Safer Clinical Systems. At its broadest level, system's thinking encompasses a large, fairly amorphous body of methods, tools and principles that consider the interrelatedness of system components and sees them as part of a common process. A system is perceived as a whole, whose elements 'hang together' because they continually affect each other over time and operate towards a common purpose.

Towards the end of Phase 1, we developed the Safety Improvement Case derived from industry Safety Cases as a means of pulling together the work done by the four organisations into an over-arching document that described the current state of safety in the particular area the organisation had worked on. This was developed further in Phase 2. The production of a Safety Case does not require the adoption of the Safer Clinical Systems Approach per se, but to write a meaningful Safety Case does require work based on the principles of the SCS approach to be undertaken.

The triangle diagram in Fig. 5.1 encapsulates the key features of Safer Clinical Systems which emerged from the experience of sites teams in Phase 1. It shows both the unique features of Safer Clinical Systems plus those aspects on which new thought or angles were generated and a reinforcement of the key underlying features impacting on any attempt to change the way we do things in the NHS. Phase 2 was designed on this basis.

See the Health Foundation website http://www.health.org.uk/areas-of-work/programmes/safer-clinical-systems/ for further information.

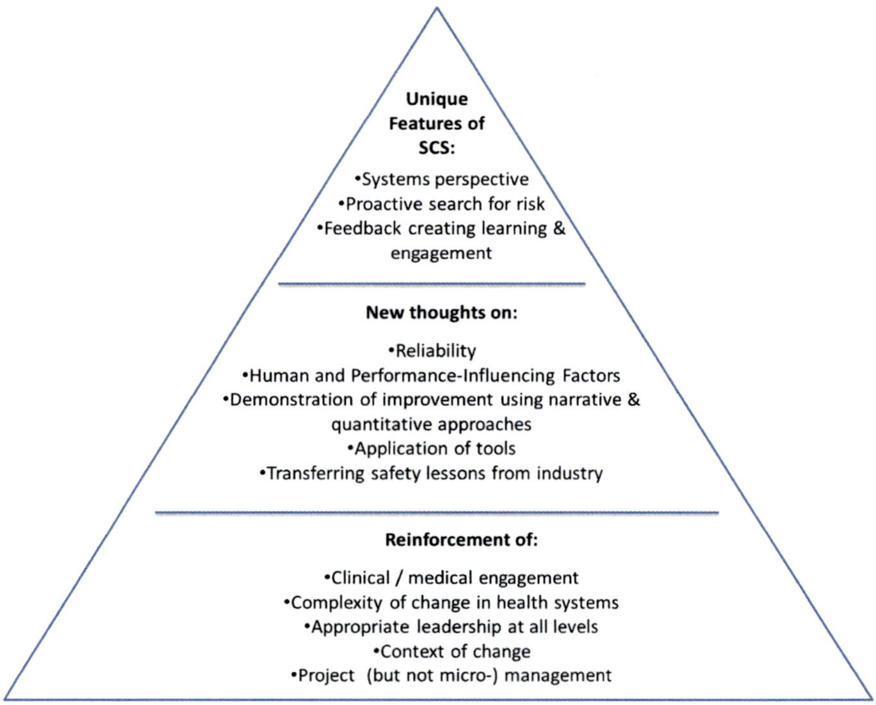

Fig. 5.1 Learning from Safer Clinical Systems Phase 1 (December 2010)

5.3 Phase 2—January 2011 to December 2013

The approach in Phase 2 was, therefore, based on 'systems thinking'—an approach to understanding how things work together, how things influence each other within a process, an organisation or any structure with a common purpose, and how to appreciate the complexity of the wider system context. There is now widespread recognition that when things go wrong for a patient, the fault rarely lies with individual practitioners, but rather with the system in which they work. By taking a systems approach, we can begin to identify the elements of a care process which may be adversely affecting safety. This means focusing on the systems that support care as well as on the care delivery itself as provided by clinicians.

The Health Foundation developed the tender for phase 2 for a core technical team to support eight selected NHS organisations to test and develop the Phase 1 concept. The Warwick Medical School team along with eight different NHS Trusts was awarded the contract. Each Trust was contracted individually. The role of the Warwick team was essential to provide consultancy support and advice to the eight Trusts.

The key to the approach adopted for the Phase 2 programme was a structured sequence of initially four steps, later expanded to five—which are described in more detail below. The fundamental requirement was that all sites would start with a rigorous diagnosis of safety issues along the patient pathway in order to understand the various factors influencing safety in that pathway. Interventions were not to be based on current assumptions about the nature and cause of patient safety issues. Specifically, this meant using selected tools to clarify the current system in terms of how it operated and what the particular hazards were within that process, together with the level of risk to patient safety they generated. Success would be demonstrated by showing that risk in the pathway had been reduced as a result of designed and targeted interventions.

The structured, sequential nature of the programme undertaken against a timescale, with review and assistance at various stages, was designed to ensure that the interventions undertaken to improve safety were the most appropriate for that pathway and that improvement was measured. Any solution must have a clear rationale based on effective diagnosis and understanding of the particular safety problem, including the existing risks as well as harm.

The overall Phase 2 programme was a complex project to both set up and to manage.

5.4 Roles

The Technical Support Team provided by Warwick Medical School, managed the overall programme and provided facilitation and training to support each of the eight sites. This was done through a combination of central training and shared learning events, virtual events, e.g. WebEx/webinars and expert technical support on site, together with a dedicated Site Facilitator with an organisational behavioural background to work with site teams acting as a critical friend, and able to channel the appropriate technical expert support on site when needed. The eight Trusts were independently contracted and accountable to The Health Foundation.

The Health Foundation team met regularly with the Support Team to review progress and the continuing development and adjustments of the programme using a system of quarterly reports and meetings on a project management basis.

There was an independent evaluation team appointed from the University of Leicester. The remit of the Evaluation Team was to focus on evaluating outcomes of the Safer Clinical Systems approach within the award holder sites [See Chap. 9 Evaluation of the SCS Approach]. The Warwick team remit was to work with the eight sites to train and support them in implementing the current version of the Safer Clinical Systems Approach and to capture learning about aspects of the approach, i.e. what works best. For example, in relation to the diagnostic and other tools used by the sites.

5.5 Timescales

The Phase 2 programme commenced in January 2011 and was originally scheduled to finish in June 2013. However, there were two extensions to the programme. The first in July 2011 was to allow more time to recruit and select the eight participating organisations and had the effect of extending the end-point of the programme to September 2013. In February 2013, it was agreed that in order to ensure an adequate period of data collection, the programme would be further extended to December 2013. This increased the contract period for sites and for the Warwick team beyond original expectations, with some consequences for staff retention during the last six months of the programme.

During the first nine months of the programme [January 2011 to September 2011], the Warwick team worked closely with the Health Foundation to refine the testing of the approach and the programme to be followed by the eight participating sites and to finalise the recruitment process for award holders and participate in recruitment and selection of award holders.

5.6 How to Build Safer Clinical Systems—A Description of the Approach

Safer Clinical Systems is a way to reduce risk and harm to patients, improve reliability and develop a better safety culture. It has **five key steps**, which we describe below. It uses tools and techniques developed from the Safer Clinical Systems programme or modified for healthcare from other safety-critical industries.

The essence of Safer Clinical Systems is to approach safety proactively, but there are also emphases on creating a culture of safety, on using human factors properly and on assessing and assuring safety to internal teams and stakeholders.

Safer Clinical Systems call for a new paradigm in understanding, designing and managing our healthcare systems to work for patient safety. To get started, here is a graphical representation of the five steps used to create Safer Clinical Systems (Fig. 5.2):

Step 1**: *****Pathway definition and context****—scope the pathway to be developed and be clear about its boundaries. Use qualitative and quantitative assessment of organisational and safety culture.*

Step 2: System diagnosis*—high-level mapping of the pathway and its linkages to wider systems. Identification of hazards; detailed human factors analysis of critical steps, risk factors and performance influencing factors; setting reliability measures and targets for outcomes and key care processes.*

Step 3: Option appraisal*—develop a shared understanding of the risks to patients in the pathway based on the outputs of Step 2. Consider options to reduce risk and build safety and test these options against outcome and practicality.*

Step 4: Planning*—define patient safety improvement objectives in risk reduction and reliability improvement; design interventions to be enacted through redesigning 'hard' systems (the things we do to care for patients) and 'soft' systems (the way we interact with patients and each other, and the way we use the hard systems).*

Step 5: System improvement*—carry out and evaluate system improvement cycles; work with teams to carry out initial system redesign followed by continuous improvement cycles; reassess hazards and risk using the same techniques employed in the diagnostic (Step 2). Use human factors interventions to support change.*

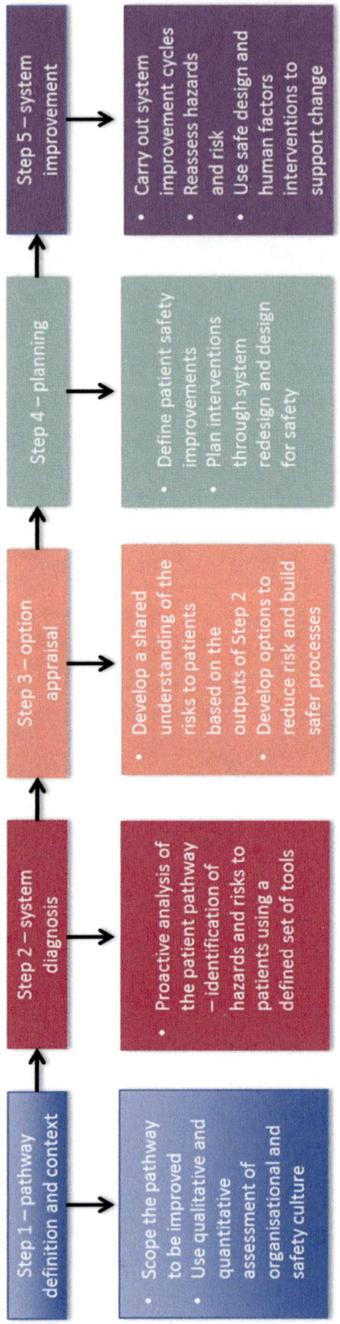

Fig. 5.2 Safer Clinical Systems overview

5.7 Safer Clinical Systems—The Five Steps

In this section, we outline each step in more detail and what you might expect to gain from each of them and provide a general guide to the tools and techniques you might apply. The five steps are set out sequentially, but you should expect there to be a degree of iteration between the steps as learning accumulates.

A robust programme structure can provide a useful backdrop for undertaking this work. A multidisciplinary team can provide vital support as 'critical friends' and develop a measurement plan to identify improvements in safety. An Executive Lead for the programme is essential to provide access to the Board, partly to ensure awareness, but most importantly to facilitate the extension and sustainability of the methodology.

Safer Clinical Systems is about changing the way things are done. It is a unique risk-based proactive approach that uses defined microsystem projects as a springboard for creating system-wide sustainable change that has an impact across the whole organisation. In approaching its introduction, it can be helpful to think in terms of two concepts—change and transition. Change involves shifts in organisational situations and processes, in ways of doing things, whereas transition is the psychological reorientation people have to experience when a significant change takes place. 'The change event may be viewed as gain but the process of transition from pre- to post-change is often experienced a loss, e.g. status, resources, moving from the known to the unknown, is that people resist, not necessarily the change itself. Intellectually, the change may be perceived as good or necessary but emotionally as negative.

Resistance is a natural, universal, inevitable human response to a change that someone else thinks is a good idea, and resisting change or improvement does not make someone bad- or narrow-minded. We have all done it and our response will be one of three things: fight, flight or freeze. So, we need to think about and work hard at, building engagement among the key stakeholders affected—those who have to help make the change happen and those that have adapted and supported the change.

The change equation is a useful tool in recognising and understanding the change from another person's point of view, before attempting to find a way to overcome any resistance and can help to shape a strategy for the process of transition (Beckhard and Harris 1987).

Dissatisfaction + Vision + Capacity + First steps > Resistance

Dissatisfaction—with the present situation

Vision—an understanding of what the change would look like

Capacity—sufficient resources to make the change happen

First steps—an appreciation of *how* the change is to be implemented

Resistance—the psychological and/or resource cost to the individual or group, of making the change.

If any of the four elements are zero, there will be insufficient impetus to overcome any resistance to change. **The biggest problem may be the assumptions [untested] we make about what people know or understand.**

5.8 Step 1—Your Pathway and Its Context

Why This Is Important?

The purpose of this preparatory step is to allow your team to become established, to begin to define the pathway, to build relationships with the staff working in the pathway and—most importantly—to develop a deep understanding of your culture and context.

Because patient care takes place within the organisational culture and context and is influenced by them, we begin with an organisational assessment. You will need to develop an understanding of both the Trust and the culture within the pathway itself. To do this, we recommend you to use two key diagnostic tools: the *Manchester Patient Safety Framework* (MaPSaF) and the *Safety Culture Index (SCI)* (National Patients Safety Agency 2006). (Further detail of each of these given later in the text).

What Do We Mean by 'A Pathway'?

The *pathway* is defined by a patient journey, not by organisational structures. As Safer Clinical Systems is based on taking a system's approach, it is necessary for the pathway being studied to extend outside any single 'microsystem' to take account of what happens before and after that might impact on patient safety. This should enable you to 'zoom out' to look at wider organisational and contextual issues that affect what happens in the microsystem rather than just looking within, to the limited subculture in a single department or team. The belief is that this will then lead to changes in those influencing factors that may inhibit change or may help the sustainability of a specific change and enable the learning to spread more widely in the whole system.

Tools and Techniques You Can Use

Existing Evidence of Safety

- Gather data from the Incident Reporting and Analysis systems where it is available.
- Refer to other sources such as safety and quality Dashboards.
- Highlight past harms and some of the factors that have been associated with them.

Manchester Patient Safety Framework (MaPSaF)

- Use the MapSaF tool to:
- Facilitate reflection on patient safety culture
- Stimulate discussion about the strengths and weaknesses of the patient safety culture
- Reveal any differences in perception between staff groups and between staff and patients/ carers
- Help understand how a more mature safety culture might look
- Help evaluate any specific interventions needed to change the patient safety culture.

The Safety Culture Index (SCI)

- Use the Safety Culture Index (Spurgeon et al. 1999) to:
- Provide a quantitative assessment of key components of organisational culture using the UK-based norms from other health organisations
- Evaluate twelve key dimensions of safety
- Determine differences between professional groups in their understanding of the patient safety culture they experience
- Signpost cultural interventions.

Your Outputs from Step 1

At the end of your Step 1, you should have a collective agreement about the pathway, the culture (especially the safety culture) within the organisation and the pathway team itself (i.e. the staff responsible for delivering patient care on this patient journey), and detailed knowledge of past harms that have occurred.

If you have used MaPSaF, you will have started a shared discussion about safety and your organisation's level of maturity, and you will know something about the areas you want to change—perhaps in your use of incident reporting, perhaps in the depth of your root cause analyses (RCAs). Finally, if you have used SCI, you will know how you compare with others across key dimensions of safety and where your improvements should be sought.

It is not essential to use both of these tools, but at least one—probably SCI—should be part of this first step. It is possible to use MaPSaF after the SCI to explore some of the areas of concern that are revealed.

5.9 Step 2—System Diagnosis

Why This Is Important?

The purpose of this next step is for the team to undertake a detailed diagnostic assessment of the pathway you have chosen and to understand in real depth the factors affecting the current safety status, both positive and negative. Out of this step will come the identification of hazards, risk, reliability and harm in each pathway, as well as specific process and contextual measures. In short, you will know where in the pathway your patients are at risk of harm, and you will begin to see why, and what can be done to prevent it.

Tools and Techniques You Can Use

Process Mapping

- Use process mapping to:
- Create a high-level description of the stages and key elements in your patient pathway
- Develop a collective understanding of the pathway and the people, departments and processes involved
- Provide a basis for detailed hazard and risk assessment to improve safety.

Failure Mode and Effects Analysis (FMEA)

Use FMEA to:

Assess the things that can go wrong for patients at each stage of the process map
Identify the most significant risks in the pathway—risk ranking
Build a collective understanding of the high-level risk profile in the pathway.

Human Factors Analysis

Use the human factors analysis to:

- Unpack high-risk elements of the process map to understand the details of patient care in the pathway
- Create a Hierarchical Task Analysis (HTA) of key tasks influencing safety
- Zoom into understand the way things go wrong
- Identify and share the human factors that introduce risk (including 'performance influencing factors')
- Understand how human error in the pathway influences risk.

Your Outputs from Step 2

Step 2 provides the foundation for improvement. Without a full system diagnosis, clinicians and managers will not be aware of the ranking of risks and therefore where their action should best be targeted. In addition, experience of using this methodology has shown that in most cases (1) team members acquire a much better understanding and appreciation of the role of others and (2) risks are revealed during this step that were not previously recognised. Many of these newly recognised risks are significant in patient care.

At the end of this step, you will have a detailed understanding of the processes and tasks taking place in your system or patient pathway. You will know something of the reliability of key elements and you will know where failures are likely and where patients are at risk. You will also know something of how humans affect the safety of the system, both positively and negatively.

Taken together with the outputs of Step 1, in which you will have assessed the context and the culture of your patient pathway, you should now be ready to look at options for change and to recognise which ones will make the greatest difference to the safety of patients.

5.10 Step 3—Option Appraisal

Why This Is Important?

By now, you will understand the context and culture, know where your patients are at risk of harm, which risks are most problematic, how humans contribute to risk and safety and what influences their performance—and you will be ready to effect change.

The purpose of this stage is to assess options for change, select preferred options and develop the *action plan* (Step 4) that will be implemented in *system improvement cycles* (Step 5). During Step 5, your changes will be tested and evaluated and may result in a revisiting of alternative options.

Tools and Techniques You Can Use

There are many tools available for option appraisal. To be successful, option appraisal must make sure that all risks are clearly identified and ranked, without exception. This is because, in practice, there can be a natural tendency to exclude from consideration options that seem too difficult to enact due to budget, human resource or culture. If key options are ruled out, then the team has a responsibility to use the organisation's governance processes to record and escalate the uncontrolled risks.

This means that option appraisal should use a two-stage process. Stage one is to rank the risks; stage two to address the feasibility, costs and return on investment. Without stage one, the truly difficult issues will not be addressed or escalated.

Use Option appraisal to:

- Identify possible interventions on the basis of the risks identified in Steps 1 and 2
- Address the risks in the pathway by rank
- Address context as well as the tasks or processes themselves
- Appraise feasibility and practicality.

We emphasise the importance of using a robust ranking of risk in carrying out option appraisal. This is important because it helps to ensure that, at the very least, the seemingly difficult problems—those that might go beyond the responsibility of the improvement team to matters of organisational culture and resource, economics or even political context—are recorded.

Your Outputs from Step 3

Options for change have to be developed by consensus. There are no foolproof methods for doing this, but the involvement of all professions and stakeholders is critical. It is quite possible that the options you develop and appraise may sometimes be ruled out by the resource limitations in your organisation, or by other practical considerations; you may have to carry out an iterative process in deciding what is practical. Your key output will include:

1. A clear description of each option—what it is, how it was derived and assumed impact/outcome
2. A statement of the criteria to be used in assessing one against another, for example,

   Degree of risk reduction expected
   Ease of implementation, i.e. technically simple
   Time required to implement, i.e. quick operational changes within pathway versus longer term strategic changes in the wider system
   Fit with trust strategy, vision, values
   Acceptability to other stakeholders
   Measurable impact on reducing risk
   Measurable impact on improving reliability
   Cost

3. Weighting each option against the criteria

Note: 'option appraisal' implies that all options are recognised and that there is an understanding of what interventions may be used to address risk. For that reason, there is considerable overlap between option appraisal and intervention planning. These two sections should be read together so that the techniques in Step 4 can inform the options in Step 3. In addition, Step 5, System Improvement, may also draw on the techniques in the next section, as even more knowledge is built around the pathway and its risks.

5.11 Step 4—Planning

Why This Is Important

Detailed planning is important for any improvement work. Having identified your key risks to manage and your options for change, you should be ready to plan your Safer Clinical Systems programme. For this, you will need to understand some of the interventions you could make.

Tools and Techniques You Can Use

Hierarchical Task Analysis

- Revisit your outputs from Step 2

Use your detailed process map or hierarchical task analysis (HTA) to:

- Identify where tasks need support or redesign for safety
- Challenge the process design—do you need additional safety steps? Can you remove any?
- Identify and manage tasks where human performance is a safety issue.

Designing for Safety

Use the tools in designing for safety to:

- Identify the most powerful way to manage a risk (*the hierarchy of control*)
- Prevent mistakes (*managing human error*)
- Use checklists correctly (*checklists*)
- Design change through consensus (*participatory design*)
- Improve staff involvement (*engaging conversations*)
- Identify drivers and resistors to change (*force field analysis*).

Your Outputs from Step 4

This step and the previous one will have been your opportunities to reflect on the findings of the system diagnosis phase and identify practical improvements. In healthcare, the urgency of the need to build safety often leads to rapid responses. This is understandable and sometimes we may have little choice. But the tallest buildings have the deepest foundations; good planning, based on detailed knowledge, will do a better job!

Step 4 will have provided you with a plan to improve patient safety in your pathway based on the proactive identification of risk. You will have pinpointed the problem areas and developed plans to eliminate, contain or minimise risks of harming patients—and you will have involved others in the process.

5.12 Step 5—System Improvement

Why This Is Important?

At this point, you will have identified and listed all the hazards in your pathway and their relative risk, identified all current mechanisms for risk control and highlighted possible interventions you may need to make by considering the nature of the underlying causes of risks. During Steps 3 and 4, you will have systematically evaluated potential interventions and have developed a planned series of interventions that you are now going to carry out. The purpose of Step 5 is to carry these out and continuously evaluate your system improvement cycles.

The key element in system improvement is **measurement**: you must be able to demonstrate progress to maintain the support and commitment of staff at all levels.

Tools and Techniques You Can Use

Measurement Plan

Use a measurement plan to:

- Set clear, quantifiable goals for safety improvements
- Ensure that your plan addresses patient outcomes as well as reliability of essential care processes
- Ensure that you have considered possible negative effects of your work
- Share information and build co-operation.

5.13 The Safety Case—(More Details and a Worked Example of Use of a Safety Care Is Given in Part II)

Use a safety case to:[1]

- Bring together all elements of your work: culture, diagnostics, risks and risk control measures, key improvement actions
- Serve as a basis for continual improvement in safety through including your measurement plan
- Provide assurance to internal and external stakeholders.

[1]Using safety cases in industry and healthcare. www.health.org.uk/safetycasesreport.

5.14 Your Outputs from Step 5

Your key outputs from Step 5 will be the changes you bring to the risk and safety of your pathway. Your impact on patient safety (through both outcomes and care process reliability) should be demonstrable and shared as widely as possible.

Further measures of culture (through the Safety Culture Index, for example) may also demonstrate changes in the attitudes, beliefs and behaviour of staff. We now move to the implementation of this SCS approach in Part II.

References

Beckhard, R., & Harris, R. T. (1987). Organisation transition: Managing complex change.

Cooke, M., Cross, S., Flanagan, H., Jarvis, R., & Spurgeon, P. (2016). *Safer Clinical Systems. A new, proactive approach to building safe healthcare systems*. A reference guide for clinicians and managers. Coventry: Warwick Medical School (Safer Clinical Systems team), University of Warwick.

National Patients Safety Agency. (2006). *Manchester patient safety framework: Facilitator guidance*. University of Manchester.

Spurgeon, P., Barwell, F., Parker, L., et al. (1999). Safety culture index. In *Organisational culture and its potential relationship to cultural risk*. London: University of Birmingham.

Part II
Implementing Safer Clinical System—Examples of SCS in Practice and Outcomes; and Next Steps to Wide Scale Dissemination

Chapter 6
Building Safer Healthcare Systems

Implementing Safer Clinical Systems Methodologies in Acute NHS Trusts

6.1 Introduction and Background

Earlier chapters provide an appraisal of the continuing issues in patient safety and challenge the existing model of patient safety, which, despite many years of work since *An Organisation with a Memory*, remains essentially reactive. In this chapter, we examine the application of the Safer Clinical Systems model—a proactive, systems-based model—on patient safety.

6.2 The Safer Clinical Systems Approach

In response to the slow progress towards safety, a proactive approach, largely modelled on safety systems from other high-risk industries and with a strong focus on human factors, was applied in a number of UK acute Trusts. The programme followed the steps described in Chap. 5 (as shown in Fig. 6.1). This chapter, the first of Part II (practical implementation) section, is aimed at documenting how these steps worked and just what was done to implement them.

This chapter describes some of the actions in four sections. Firstly, we look at the **organisational context** in more detail: what are the existing learning systems of the Trusts; how can the local culture be measured and how it is experienced by staff? This section will add to cultural analyses conducted through the Safety Culture Index (SCI) described elsewhere in this book and describe in more detail the significant problems with a purely reactive approach to safety. Secondly, we provide examples of the key second phase of the Safer Clinical Systems programme—**System Diagnosis**.

In the third section, we describe some of the **interventions** that arose as a result of system diagnosis, with an emphasis on the principles involved, and ask what these have achieved in terms of reducing risk for patients. Finally, we examine and exemplify a holistic approach to safety in systems, frequently applied in other high-risk industries, known as a **Safety Case**. This practical tool, which also has important

P. Spurgeon et al., *Building Safer Healthcare Systems*,
https://doi.org/10.1007/978-3-030-18244-1_6

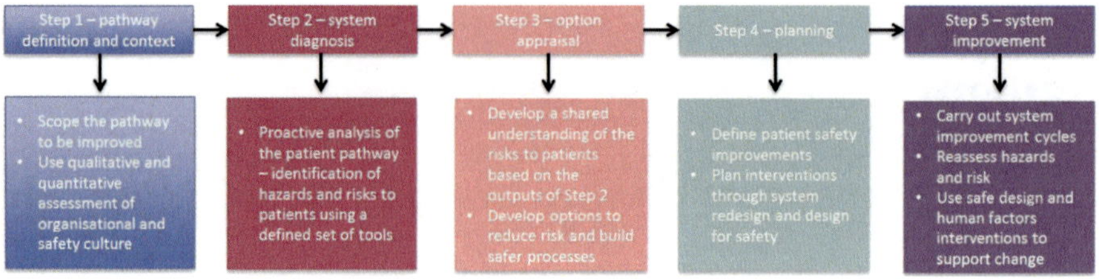

Fig. 6.1 Overview of SCS programme

implications for the fraught area of regulation in healthcare, brings together the knowledge—including the tacit knowledge of clinicians—of safety hazards, risks and overall safety in a particular pathway or area and can serve as a working system for building Safe Clinical Systems.

6.3 The Organisational Context

What is the reality of organisational culture and organisational learning in the NHS?

6.4 MaPSaF

Understanding the existing culture is an essential element of the diagnostic phase of Safer Clinical Systems and has a wider utility in judging the effectiveness of the existing systems of safety management, which are largely reactive. Several cultural assessment tools were applied during diagnostic phases of Safer Clinical Systems, which, taken together, paint an accurate and detailed picture of the reality of NHS culture.

In addition to a detailed and formal assessment of safety culture using the Safety Culture Index (Chap. 7), each site on the Safer Clinical Systems programme carried out a cultural evaluation through workshops using the Manchester Patient Safety Framework (MaPSaF). This tool is frequently used as much as a cultural intervention as a cultural measure, since it supports local clinical teams in reflecting on safety systems and how they are locally applied. However, these reflections can provide a useful measure of local safety culture—especially useful when a new approach to safety is begun.

The tool uses a development of Westrum's organisational maturity framework and places each Trust on a scale of increasing organisational maturity with respect to safety culture, from *pathological* at the lowest level, through *bureaucratic* to *generative* at the highest level of maturity, as illustrated in Fig. 6.2.

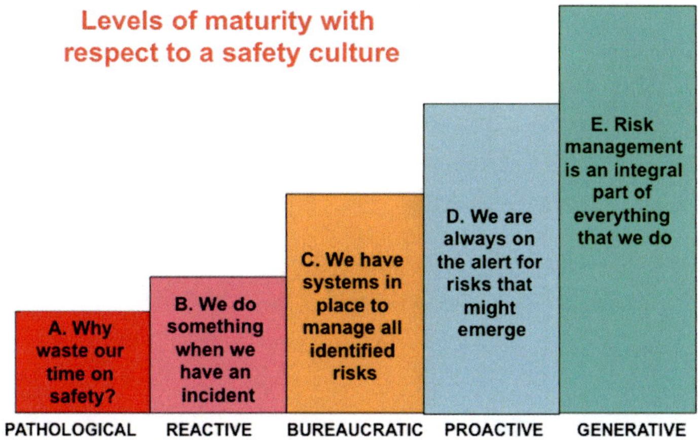

Fig. 6.2 MaPSaF framework

There are several dimensions in MaPSaF, each of which can be considered separately. Participants discuss and form a consensus view on safety culture in each of the dimensions, based on illustrative statements. To illustrate this, participants identifying their safety culture as 'bureaucratic' in the dimension 'priority given to patient safety' would agree with the statement:

> Safety has a fairly high priority and there are numerous systems (including those integrating the patient perspective) in place to protect it. However, these systems are not widely disseminated to staff or reviewed. They also tend to lack the flexibility to respond to unforeseen events and fail to capture the complexity of the issues involved. Responsibility for risk management is invested in a single individual who does not integrate it within the wider organisation. It is an imposed culture.

Placing the organisation at the 'generative' level, however, participants would agree with:

> Safety is the top priority in the organisation, and responsibility for safety is seen as being part of everyone's role including patients and the public.
>
> Staff constantly assess risks and look for potential improvements. Patient safety is a high profile issue throughout the organisation and is embedded in the activities of all staff, from the Board/senior managers through to healthcare teams who have day-to-day contact with patients, including support staff. Patient involvement in, and review of, patient safety issues is well established.

In our MaPSaF sessions, we considered six dimensions of safety culture:

- Priority given to patient safety
- Systems errors in individual responsibility
- Recording incidents and best practice
- Learning and effecting change
- Communication about safety issues
- Team working.

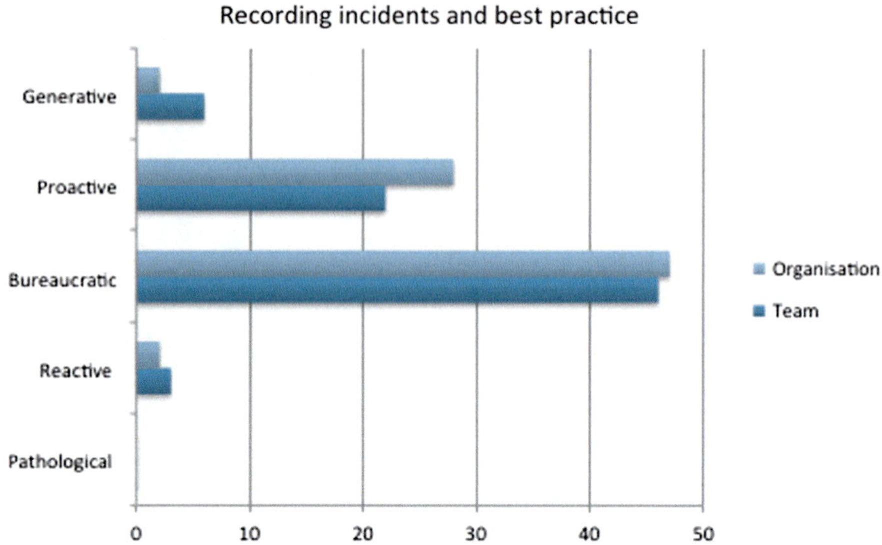

Fig. 6.3 Responses to MaPSaF framework

In this evaluation, it was rare to find an organisation that teams placed at the more mature end of the framework. Most teams assessed maturity with regard to safety as *bureaucratic* or sometimes *proactive.* This was especially the case with the dimensions relating to communications and learning from incidents. Figure 6.3 illustrates overall responses, based on approximately 100 MaPSaF participants, for the dimension *recording incidents and best practice*.

In the workshop discussions, some key points about safety culture were underlined by participants, which are relevant to the development of safety culture and serve to illustrate the starting point for the introduction of a proactive approach. A common theme was the perception of a disconnection between senior management and the 'sharp end' of clinical practice. Related to this was a sense of a gap between the stated policies of the organisations and the practices as experienced by clinical staff and safety or governance managers who made up the SCS teams at Trust level.

More than anything else, teams identified a lack of transparency in organisations and the continued existence of a culture which blamed individuals rather than systems for safety failures. A culture of blame is a significant factor in suppressing the ability of the organisation to learn and create safety for patients, and there is probably no organisation in the healthcare sector without a stated policy of either no-blame or 'just' culture. In such a culture, incidents and risks (often revealed by 'near misses' where no harm to patients occurs) are to be investigated openly, even when human error is involved, in the expectation of organisation change. This process is at the heart of safety management in the NHS; the degree to which this takes place is not only relevant to the utility of the reactive system of safety management but is a key cultural attribute which must be understood when introducing proactive methodology; a culture which believes that mistakes are punishable failures may not be expected to respond positively to safety initiatives generally, for example.

6.5 Reporting and Learning

A more detailed evaluation of the reactive culture was possible in a single Trust, where a survey of reporting and learning culture was carried out as part of the diagnostic phase. In this Trust, which was under detailed regulatory review as a result of patient mortality as measured by hospital standardised mortality rate (HSMR) had received several inspections and subsequent internal initiatives entirely centred around incident reviews and mortality reviews. Results of these initiatives, both cultural and in terms of creating safety, were revealing—and cast the reality of reactive safety management in healthcare in a harsh light.

This survey instrument, which has been applied in all health boards of NHS Scotland (NHS Scotland 2007) and used in other safety-critical industries, is based on a model of organisational learning and which of several key organisational attributes may be crucial.

Put simply, to be a good learning organisation, in line with NHS policy, a culture is required where staff have a good understanding of why things go wrong, where they have a high motivation to report and investigate, and where the systems used to support this show good usability and effectiveness. The survey addresses these issues through five key dimensions:

- Attitudes and beliefs about error
- Consequences of admitting to a mistake
- Sharing of experience
- Organisational response to problems
- Characteristics of the reporting system.

To have a good learning organisation, staff and management firstly need to understand that human error is an inevitable consequence of the way we work and that in most cases individual blame is inappropriate and merely serves to suppress reporting by creating a culture of fear.

Secondly, the organisation needs to respond appropriately to errors through sharing experience, improving systems and practices and—most importantly—ensuring that people who honestly report error are not unfairly treated. Finally, the system itself has to be functional: it must be easy and effective to report an error and to analyse an incident. Good scores on these dimensions are essential if the reactive safety management system in universal use in Trusts can ever be considered effective.

Overall scores in these dimensions are shown in Fig. 6.4, for both the Trust where the survey was carried out and for NHS comparators. The data are presented on a Likert scale of 1–5, where a strong system would be represented by a score of 4 or above.

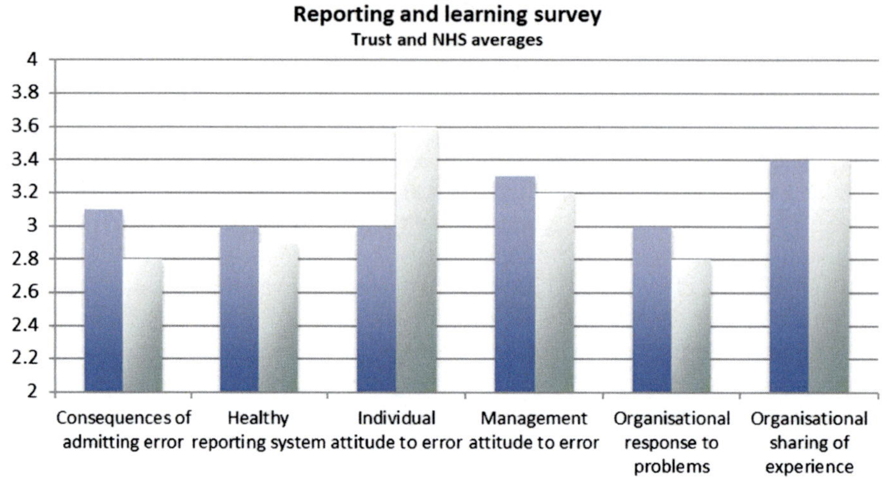

Fig. 6.4 Reporting and learning survey

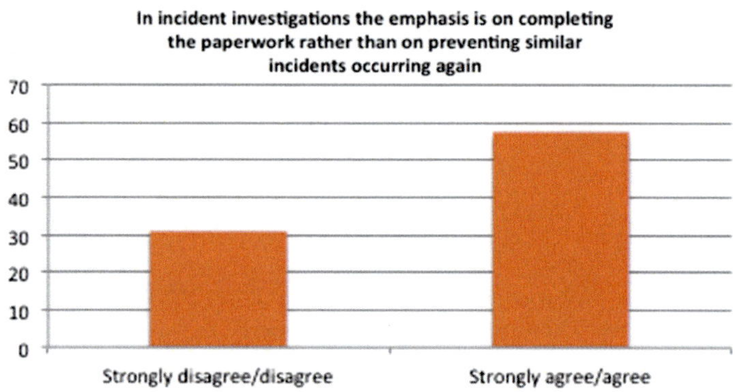

Fig. 6.5 Incident investigation responses

Neither the Trust sampled here nor previous measures from the NHS organisations achieve 'healthy' scores in this survey. That this reflects a dysfunctional learning system seems clear, but some of the responses to the 60 individual questions which make up the survey are still more revealing. For example, the very basis of reactive learning—the reporting of adverse incidents—is questionable, with most respondents of the 98 respondents in this Trust, who were drawn exclusively from frontline staff, believing the process to be primarily bureaucratic; Fig. 6.5.

Worse, investigations following a safety incident focus on individual blame and frequently ignore the contextual or system factors that may have contributed to the event; Fig. 6.6.

These surveys are important because the learning culture in the healthcare sector is *reactive by design* and it is upon this that the safety of patients currently depends.

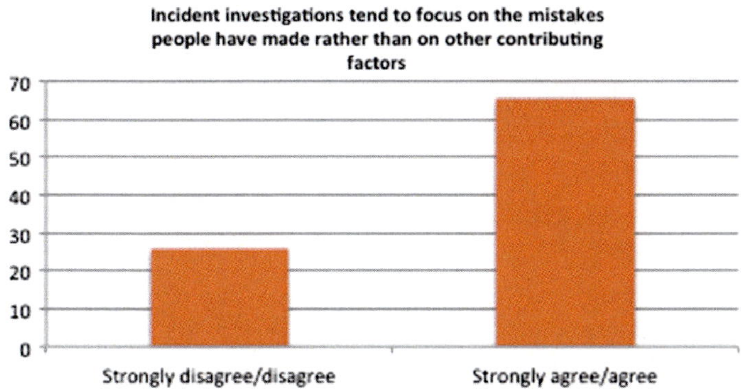

Fig. 6.6 Focus on mistakes versus other factors

We would make two major points on the basis of these diagnostic tools. Firstly, though enormous effort is invested in this reactive system, it simply *does not work as intended*. A new approach is needed, in addition to the 'repair' of the existing culture of bureaucracy and blame.

Secondly, in a reactive system, harm has to occur before learning takes place or risks to patients are addressed. A culture of recrimination and blame is therefore almost impossible to avoid. In contrast, where risks are considered and addressed proactively, no specific incidents are in the spotlight, no harm has yet occurred, and no blame is possible. A proactive systems-based approach may be the only practical way of creating an antidote to the persistent and destructive culture of blame in healthcare.

6.6 Developing Safer Clinical Systems

Assessment of safety culture, derived from all the tools used during the Safer Clinical Systems programme, provides a bleak but recognisable picture. At the time that the programme began, staff in these Trusts believed that the organisational priority given to patient safety was mediocre. The participating teams themselves had a strong commitment to and interest in safety, however, and a sharp appreciation of human error and the dangers inherent in real-life systems. Despite a near-complete reliance by the organisations on reactive learning, respondents in surveys saw the reporting and analysis symptoms as difficult to use, bureaucratic and so focused on individual blame as to usually ignore the systems and human factors that dominate events leading to patient harm.

How participants' understanding of safety and safety management changed as they contributed to safety analysis, improvement and the construction of a holistic commentary on system safety, the *Safety Case*, is described later. An essential element of this change, which was reflected in changes in formal cultural metrics (Chap. 7), was the detailed analysis and 'diagnosis' of the system the teams had chosen to work on. In this section, we describe the key tools that were applied and illustrate some of the outcomes from the analyses.

6.7 Diagnosis—Rationale and Overview

The overall objective of these analyses was to develop a shared understanding of the systems in the multidisciplinary teams that participated. Specifically, the diagnostic approach supported by the teams aimed to:

- Map the system in appropriate detail
- Identify steps or tasks within the system where risk was present
- Evaluate the existing risk control measures
- Prioritise those risks to patients that need management.

This brief outline requires some immediate clarification. System mapping, whether through thorough process mapping or more rigorous task analysis, involves the decomposition of tasks into those subtasks that form the steps of the process. For example, the task 'assess the patient' usually includes six subtasks relating to physiological observations and whatever specific additional observations are necessary for the patient's specific condition. This deconstruction can theoretically be taken to any level of detail, so what is an appropriate level of detail? The usual answer is pragmatic—there is a need to identify where in the subtasks the risk is present so that it can be addressed; deconstructing tasks which never go wrong is unnecessary. In our example of assessing the patient, most tasks are reliable and some are automated, but manually checking respiratory rate is sometimes carried out too quickly and errors can lead to underestimating the vulnerability of the patient through the hospital's early warning score systems. Tasks which are error-prone or unreliable and critical to patient safety may require more detailed deconstruction to understand where and why failures are located.

We should also be clear what is meant by 'risk' in this context? Formally, we define risk as: *the combination (usually multiplicative) of the likelihood of occurrence of a hazard and the severity of the consequences of a hazard (frequency x severity).* What is a 'hazard' then, and why is different to a risk? In a clinical context, we could define a hazard as 'something that could lead to harm'. A cable lying across a busy corridor is a hazard. The risk associated with the hazard is the chance of someone falling over the cable multiplied by the harm that might be done.

In practice, this really means the chance of something going wrong for patients. In many industries, the estimation of risk may be precise and quantitative—how likely is this valve to fail or this pilot to press the wrong button?—but in healthcare, our estimated risk is usually derived through a highly subjective process using fairly crude measures such as 'highly likely' or 'moderate harm'. Despite this, we believe that these methods and estimates have real value. This is especially true when the knowledge and estimation of risk are carried out by teams of involved professionals through consensus. Who would *not* wish to know what the professionals involved in caring for patients thought about hazard and risk?

When a system—a patient pathway, for example, or a defined physical area such as a ward or a theatre—is mapped and risks are identified using the formal systems we describe here there are two additional outcomes. Because these analyses are best conducted as group exercises, it is likely that the tacit knowledge of teams will be made explicit. This surfacing and sharing of knowledge has its own value, but it also leads to the ranking of risks on a simple scale. This identifies the most important risks to tackle and also enables the consideration of to what degree there are existing risk control measures. For example, a powerful drug might represent a serious hazard, but measures may be in place to be sure that its dose is reliably checked at several points in the system. The 'residual' risk of the drug is therefore low in this case.

A final point relates to the estimation of risk in clinical systems is that, regardless of whether the numbers or rankings represent an objective reality, they mean something to the teams involved. The highlighting of risk and, very importantly, their relative ranking informs clinical teams bent on improving patient safety where their attention is best applied. Safety depends fundamentally upon the management of risk.

6.8 Tools and Techniques

In the Safer Clinical Systems programme, we carried out a depth of system analysis that is unusual in healthcare—though it is common in certain other safety-critical industries, and we drew heavily on those industries in designing the programme. Though several techniques were used, the most successful in highlighting risks to patients in this programme were *Process Mapping, Hierarchical Task Analysis, and Failure Mode and Effects Analysis* (Chap. 9 provides more information on the perceived utility of techniques and sources to identify risks to patients).

As outlined above, the thrust of these analyses is to map the elements of the system to a level of detail that enables risk to be identified and be localised, then to evaluate the strength of risk control measures and so come to a judgement about 'residual' or inadequately controlled risks to patients. Subsequent steps in the programme then address options for intervention.

We provide brief descriptions of the main tools and techniques below. An important element in them all, however, is the use of multidisciplinary groups. Constant themes in safety and quality improvement are the need to support communication outside of the 'silos' of individual professions such as nursing and medicine and the need to unlock and utilise the tacit knowledge of all staff involved in a system or pathway. A vital result of this process has been found to be the identification of risks to patients not previously recognised. A group—probably any group—of clinicians considering a system would be able to offer an immediate appreciation of the key risks to patients as they perceive them; in consensus-based, systematic analysis, however, 'hidden' risks, often significant, have been identified. But we believe it to be an essential element of Safer Clinical Systems in practice to ensure the full inclusion of all professions in the diagnosis of system safety. Detailed guides for these tools and techniques are available from many sources and have been collected in a reference guide developed by the Safer Clinical Systems support team (ref).

6.9 Process Mapping

Process mapping is perhaps the most familiar of the tools and techniques used in building safer healthcare systems. It produces a visual representation of a system as an aid to understanding, sharing information and formulating ideas for change.

Process mapping provides:

- A visual, concrete description of the pathway
- Focus on the patient and the patient voice
- Clarification of complex processes
- Understanding of the links with other systems and processes
- Knowledge of key steps and which steps add value to the pathway
- Shared understanding among the team
- Opportunity to involve and engage staff
- A launch point for identifying hazards and risks.

A process map, as a first step to understanding the sometimes complex and interlocking processes in healthcare, provides a high-level representation of the system, which then serves as a starting point in identifying risk.

6.10 Failure Mode and Effects Analysis (FMEA)

FMEA is a systematic analysis of a process to identify the possible ways it might fail. FMEA is used to examine the results of failures and the possible causes. FMEA originated in high-risk industries and has been employed in sectors such as automotive, aviation and railways for many years. More recently, it has been adopted in healthcare and, though it has received some criticism, largely on the grounds of the consistency of its results, it provides a methodology to build a shared perspective on risks to patients, a way to evaluate the relative magnitude of each risk and some insight into factors that may cause or contribute to risk and harm.

6.11 Hierarchical Task Analysis (HTA)

Process mapping and FMEA are valuable tools and relatively accessible in use to clinical staff without formal training in risk and safety. Often, though, where a 'box' or a task in a process map is identified as having associated risks, we need to carry out a more detailed analysis to really understand how things may go wrong and how to prevent, recover from or mitigate the failure. HTA is a goal-oriented analytical technique which can be used to deconstruct tasks into their component subtasks, assess the potential for failure in each component and evaluate both contextual factors, often called 'performance influencing factors' and the contribution of human error. Though HTA requires technical ability in more depth than more familiar techniques, its ability to deconstruct complex tasks provides real insight into how systems really work and what people actually do. Like process mapping, it can also be used as a launch pad for FMEA or other risk evaluation methods. In the redesign of systems, HTA can also be used to set the goals and plans of a system or process from scratch.

6.12 System Diagnosis and Building Safety

Risk Evaluations

In all, the Safer Clinical Systems approach has been applied in nine Trusts to date. In each Trust, the diagnostic phase was conducted by multidisciplinary teams with a strong commitment to a proactive approach to patient safety and the process began with a detailed investigation of risk in the pathway they chose to work on. They used initial process mapping to clearly define the scope of the programme and the connections of their pathway with other systems, and to develop an early view of the key areas of risk as they perceived them.

These analyses can be complex and detailed, with task analyses and FMEA in particular resulting in lengthy but necessary qualitative descriptions of risk areas and ranking of embedded risk which we will not present fully here. However, the scope and summary of system diagnosis can be simply illustrated below. Figure 6.7 illustrates the starting point in a Trust developing a safe surgical pathway, and Fig. 6.8, the proactive judgements of risk arrived at.

Following this, a detailed HTA took place with a view to identifying more closely the tasks involved and the performance influencing factors contributing to risk.

This analysis provides a consensus view of risk. It enabled the setting of clear goals for improvement, both through the management of process reliability and through the redesign of systems. Importantly, these measures chosen by the team to assess safety improvement related directly to the analysis of risk.

A number of Trusts chose to address medication safety specifically in their safety programmes and applied HTA in detail to the medication pathway. Figure 6.9 illustrates the goal-based approach of HTA and shows the high-level task breakdown, which was then used in further deconstruction and risk analysis. Each of the 'boxes'—representing the main goals or tasks in the medication pathway—was then broken down into subtasks, which were then in turn evaluated for risk and error potential.

From these task analyses, clinical teams were then able to identify risks, rank them and begin to assess causal factors, factors which influence task performance,

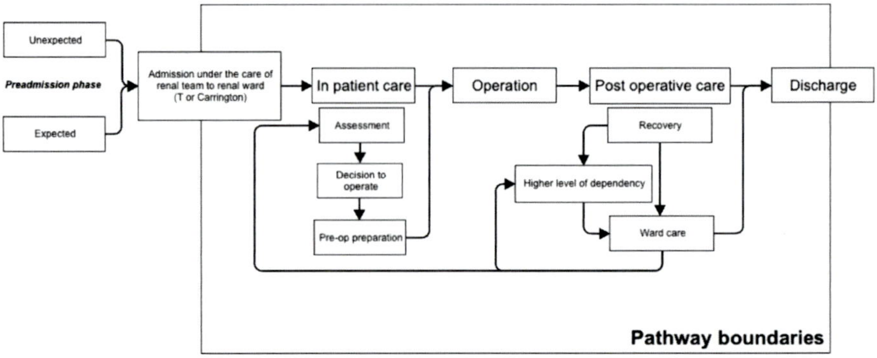

Fig. 6.7 High-level process map

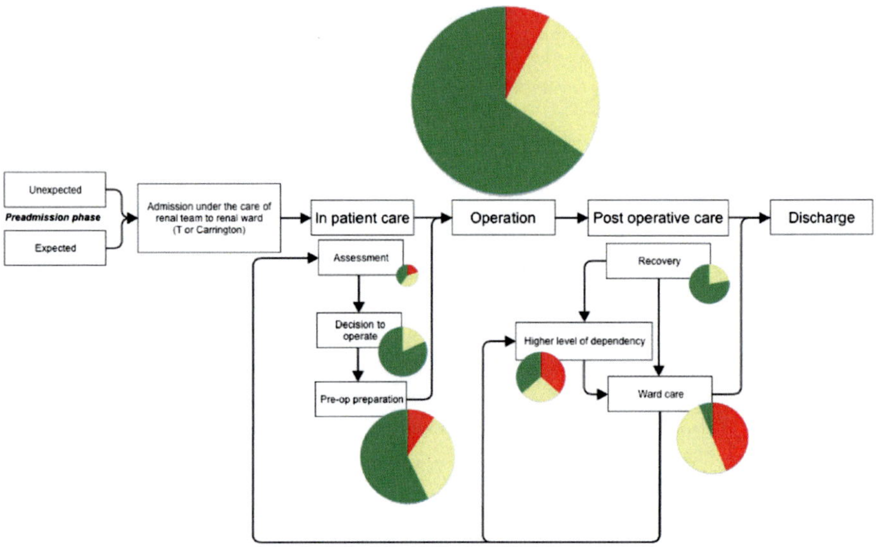

Fig. 6.8 Risk assessments

and develop measures for improvement. An example of part of this analysis in this case including an examination of performance influencing factors follows in Fig. 6.9. This illustrates a decomposition of goal 1 (reconcile patient medications) and goal 2 (prescribe required medications) into the subordinate tasks required to fulfil these goals. These subordinate tasks are numbered in the full analysis, in the 'ID' column and the risk evaluation is colour coded as high, medium or low.

Following this analysis, both system redesigns and process reliability metrics were introduced. Section x, describing the related safety case, provides some further detail—but we would emphasise that the key points are the identification of risk

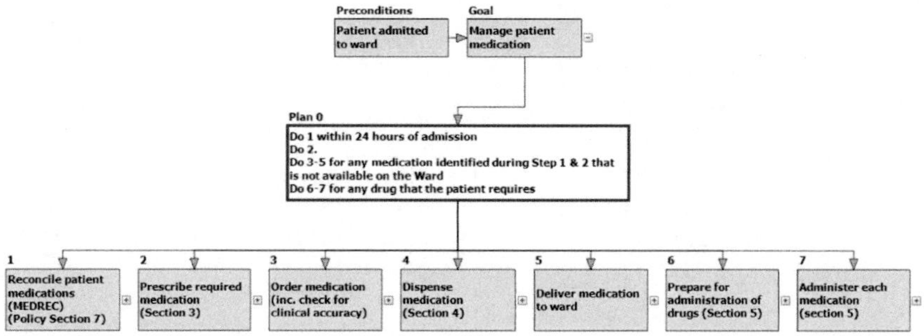

Fig. 6.9 High-level HTA of medicines management pathway

through consensus, the ranking of risk and the clear link between risk and intervention.

This particular example was ambitiously targeted at the entire medicines management pathway. Further consultation led to the analysis of specific, high-risk elements in medicines management such as the interaction with treating sepsis, where similar proactive dives into risk, and safety led to system changes (Table 6.1).

6.13 Option Appraisal and Improvement

Central to this systems-based approach to safety is an appreciation of risk. Trusts working to improve the future safety of patients tried to base their approach on the hazards and consequent risks identified in the diagnostic phase. Clearly, any proactive safety interventions should be linked to the risks identified, even when risk identification and appraisal in a complex, sociotechnical system necessarily include the subjective contributions of clinicians or consensus groups of staff. This subjectivity in risk appraisal is not a bad thing. As is the case in any safety-critical system, including healthcare, who would not wish to know how the skilled people working within the system view risks to patients? Arguably, this is the best information we can obtain given the constraints and complexity involved.

In response to these diagnoses, the chief task of Trusts in thinking about options and improvements has to be the management of the risks uncovered. As an example, Fig. 6.10 illustrates the link between risks and the interventions designed in response in an acute Trust working to create safety in the treatment of vulnerable child.

This close relationship between the diagnostic phase, where risks were identified and ranked by the clinical teams, illustrates the systems approach and the ability of the teams to begin to manage systems risk proactively—and independently of past harm events.

Table 6.1 Risk reduction per Trust

ID	Description	Role	Risk	Factors affecting performance/issues
1	Reconcile patient medications *within 24 h of admission*:			
1.1	Confirm the existing patient medication	Pharmacist, technician or doctor	H	Availability and use of protocols: It is not always possible to complete reconciliation out of hours (e.g. not able to contact GP surgery) Team structure: Nominally, this is a shared responsibility. In practice, pharmacy may take responsibility (e.g. by rechecking reconciliation performed by doctors). Processes for ensuring this is completed (e.g. if a doctor is unable to satisfactorily complete reconciliation and night) were unclear. Because this is informal, it may lead to failures (e.g. not fully completing, in expectation that it will be reviewed by pharmacy) Staffing levels and skills mix: As pharmacy staff are only available during the day and in the week, the (informal) pharmacy assumption of responsibility (see above) may be less useful out of hours Condition of packaging (design, availability and maintenance): Packages patients bring in may be in poor condition or not present (e.g. a strip of pills without box)
1.2	Record details of patient medication on drug history	Pharmacist, technician or doctor	H	Availability and use of protocols: The 'emergency clerking proforma' is 13 pages long. Fully completing this and transcribing to the drug chart (or IPP) may take half an hour to an hour. As well being a source of potential failures (i.e. in the transcription process), this may be more time than a doctor can usefully spare (as manifested in the so-called lazy clerking, where a doctor writes 'See drug chart' on the clerking document, or using the previous admission history) −ve staffing levels/skills mix: Doctors' levels of detail may be less than pharmacy (because of expectation that pharmacy will review/complete?) −ve availability and use of protocols: There may be some situations where there are well-known interactions where a technician might cross-off a dose for a couple of days. There are no hard and fast rules for this

(continued)

Table 6.1 (continued)

ID	Description	Role	Risk	Factors affecting performance/issues
1.3	Check that medicines, doses and frequencies are correct	Pharmacist, technician or doctor	H	
1.4	Identify medicine requirements	Pharmacist, technician or doctor	L	
1.5	*If full reconciliation has not been possible,* highlight the need for verification by pharmacy team within 24 h	Pharmacist, technician or doctor	M	
2	Prescribe required medication:			
2.1	Review drug history	Doctor	L	
2.2	Assess patient	Doctor	L	
2.3	Prescribe identified medication identified:	Doctor/other prescriber		
2.3.1	Ensure that patient name, NHS number and date of birth are present on each prescription	Doctor/other prescriber	L	
2.3.2	Check for allergy status		L	
2.3.3	Ensure that the admission details are complete (consultant, hospital site, ward, date of admission)	Doctor/other prescriber	M	
2.3.4	Add any other required information	Doctor/other prescriber	L	
2.3.5	Add medicine details:			
	2.3.5.1 Write drug name clearly	Doctor/other prescriber	H	Written communication: This depends heavily on the doctor's handwriting. There may also be other issues such as using brand names or Latin names. The introduction of stamps for the doctor's name and GMC number has been helpful in recovery (see 2.3.7)

(continued)

Table 6.1 (continued)

ID	Description	Role	Risk	Factors affecting performance/issues
	2.3.5.2 Specify form (if not tablet or capsule)	Doctor/other prescriber	L	
	2.3.5.3 Specify strength (for any medicine such as a cream where there are different strengths)	Doctor/other prescriber	M	
	2.3.5.4 Specify dose (ideally single)	Doctor/other prescriber	L	
	2.3.5.5 Specify route of administration	Doctor/other prescriber	L	
	2.3.5.6 Specify times of administration	Doctor/other prescriber	M	
	2.3.5.7 Specify start date	Doctor/other prescriber	L	
	2.3.5.8 Specify end date	Doctor/other prescriber	M	
2.3.6	Sign the prescription (not initials)	Doctor/other prescriber	L	
2.3.7	Add a contact or bleep number	Doctor/other prescriber	M	Written communication: Feedback was that the completion of this step varies in reliability; this makes clarifying queries about medication more difficult. There are also specific issues such as a doctor listing their on-call bleep (which is passed on at the end of their shift) rather than their own personal bleep (in effect making them uncontactable after the bleep has been passed on)
3	Order medication *for any medication identified during Steps 1 and 2 that is not available on the ward*:			
3.1	*If ordered from ward (during pharmacist round),* enter prescription details on ward console/tablet	Pharmacy techni-cian/ward pharmacist	L	

(continued)

Table 6.1 (continued)

ID	Description	Role	Risk	Factors affecting performance/issues
3.2	Complete independent medication accuracy check against prescription (and for appropriateness)	Ward pharmacist	L	
3.3	*If medication is required when a ward pharmacist is not available,* take inpatient prescription (IPP) to pharmacy	Nurse or student nurse	L	
3.4	Complete independent medication accuracy check against prescription (and for appropriateness)	Desk pharmacist	M	
3.5	Put IPP in dispensing tray	Desk pharmacist	L	
4	Dispense medication:			
4.1	*If drug has been ordered manually (i.e. paper IPP sent to pharmacy),* enter patient details and requested drug(s) in pharmacy computer	Band 3+	L	
4.2	Print labels for medication		L	
4.3	Pick medication	Band 3+	M	
4.4	Label medication	Band 3+	L	
4.5	Complete independent medication accuracy check against prescription (and for appropriateness)	Band 5/pharma-cist	L	
4.6	Leave medication in plastic bag in ward pigeonhole	Band 5/pharma-cist	L	
5	Deliver medication to ward:			

(continued)

Table 6.1 (continued)

ID	Description	Role	Risk	Factors affecting performance/issues
5.1	Bag up drugs for ward from pigeonhole	Band 3+	L	
5.2	*If drugs are controlled,* sign for drugs	Porter	L	
5.3	Take drugs to ward	Porter	L	
6	Prepare for administration of drugs:			Note: Specific forms of administration (e.g. IV drugs) have not been explicitly addressed in this analysis. These require a more detailed analysis to fully understand factors affecting performance
6.1	Prepare drug trolley	Nurse	L	
6.2	Verify patient ID against drug chart:			
6.2.1	Ask patient name	Nurse	H	Availability and use of protocols: Protocols may be undermined (understandably) by familiarity with patients, and misunderstanding of purpose of protocol. For example, one participant indicated that they check the name of a patient only once a day (unless giving a controlled drug, or if the patient cannot communicate), as they have already been 'identified'
6.2.2	Ask DOB	Nurse	H	
6.2.3	Check nameband	Nurse	H	−ve PATIENT CONDITION: Patients will frequently remove bands—particularly patients with complex needs e.g. Dementia
6.3	Check patient allergy status:			
6.3.1	Review allergy details on drug chart		M	
6.3.2	Check for patient allergy wristband	Nurse	M	
6.3.3	*if Patient is able to communicate* confirm allergy status with patient	Nurse	M	
6.4	Review drug chart to confirm patient medication needs:			
6.4.1	Ensure that the prescription is consistent with care plan	Nurse	M	

(continued)

Table 6.1 (continued)

ID	Description	Role	Risk	Factors affecting performance/issues
6.4.2	Establish medication to be given	Nurse	M	
6.4.3	Establish route by which medicine is to be given	Nurse	L	
6.4.4	Establish whether the drug needs to be given during this round (timing)	Nurse	L	
6.4.5	*If drugs are to be administered by IV* Request second check that medicines are appropriate for prescription/patient	Nurse	L	
6.5	Check for contraindications:			
6.5.1	Ensure patient's weight is stated	Nurse	H	Clarity of responsibility (team structure): This is not always recorded on drug chart, nobody has specific responsibility for recording weight, so it is an item that may be missed Duplication of information (written communication): There is a lot of information on the drug chart of which weight is just one, and weight is included in other locations Knowledge and skills: Reliance on nurse experience to know drugs which have adverse outcomes if administered when patient weight not suitable
6.5.2	Check observations	Nurse	M	
6.5.3	Check relevant tests	Nurse	H	Availability and accuracy of test results/supervision: This information is managed primarily by doctors and is kept separately, so nurses have to seek it out specifically. Moreover, doctors do not always update original prescription to take account of pharmacist observations/notes. Taken together, this means nurses may be unlikely to check test results unless some other issue prompts them to do so

(continued)

Table 6.1 (continued)

ID	Description	Role	Risk	Factors affecting performance/issues
6.5.4	Determine suitability of medication based on observations and test review	Nurse	H	Knowledge and skills: Establishing contraindications can be difficult; it requires a degree of knowledge, and nurses may find it difficult to do, particularly given time pressure Staffing levels and skills mix: Staffing levels (inc. agency staff) means that nurses will be moved from ward to ward and may not have sufficient knowledge to make these decisions—knowledge that they would otherwise have acquired over time from working on one ward Task design (time pressure): Nurses are under pressure to complete round as quickly as possible; this may undermine a desire to fully check for possible contraindications
6.6	Select medication:			
6.6.1	Select from trolley	Nurse	H	Task design: Trolley design and frequent changes in medicine packaging increase the probability of selection failures at this point. Some medical packaging is similar and vulnerable to right action/wrong object selections. There is often poor segregation of medicines in the trolley Task design (task variability): There are many different ways this could be approached (e.g. one patient at a time, all oral drugs first, all injections first). Some methods may increase probability of omissions and other failures Staffing levels and skills mix (distractions): Distractions are frequent and disruptive when attempting to complete drug rounds. There is a system in place for the use of a do not disturb bib; however, it was suggested that this is of variable value. Sometimes, there are insufficient staff numbers to enable an individual to go about the drug round undisturbed Design, availability and maintenance of equipment—Trolley layout is poor, with frequently inadequate segregation of medicines
6.6.2	Select from bedside cabinet	Nurse	L	
6.6.3	*If required drugs not in trolley or cabinet,* select from ward stock	Nurse	L	

(continued)

Table 6.1 (continued)

ID	Description	Role	Risk	Factors affecting performance/issues
6.6.4	*If drugs are in ward fridge,* select from fridge	Nurse	M	
6.6.5	*If drugs are controlled,* select from controlled drug cabinet	Nurse	L	
6.6.6	*If drugs are not available,* request order from pharmacy	Nurse		
6.6.7	Establish the dose	Nurse	H	Staffing levels and skills mix (distractions): Distractions are frequent and disruptive when attempting to complete drug rounds. This may be a particular issue when calculating doses Task design (time pressure): Nurse will be under pressure to complete round as quickly as possible Task design (informal practices): Pharmacist, if they have time, will tend to write this information in, which is of great help to nurses
6.6.8	Dispense medication into pot for patient	Nurse	L	
6.7	Discuss additional medication requirements with patient	Nurse	L	
7	Administer each medication			
7.1	Give medication	Nurse	L	
7.2	Observe medication taken	Nurse	H	Task design (time pressure): Individual patients may struggle with taking drugs; therefore, this step may take some time. There may be a temptation to leave a patient with the drug and return to them later
7.3	Sign medication chart for each patient	Nurse	M	

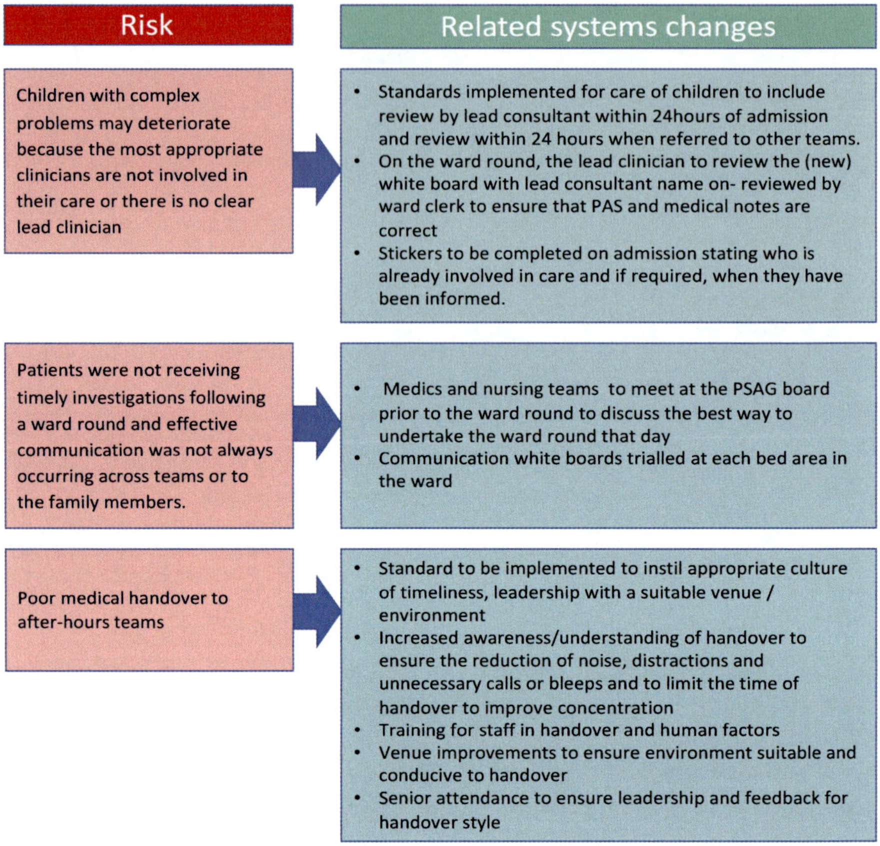

Fig. 6.10 Intervention options related to systems risks

6.14 Design of Interventions

As a general experience, these appraisals of options and improvement plans lead to two main classes of intervention. Understanding of the system through process mapping and HTA identifies what processes must take place to accomplish the goals of safe patient care and where the risks are present. The options therefore tend to unfold as *redesigning* the system to minimise or eliminate risk, and the management of *key processes*. The interventions described above, for example, are largely changes in the way clinical work was carried out. The success or failure of such interventions is best judged on that basis, that is to say, was the overall systems risk reduced? The following section, *Uncovering Risk*, provides an overall review of risk reduction through this programme.

In other Trusts, a wide range of systems-based interventions were arrived at through the option appraisal and improvement elements of the Safer Clinical Systems approach. These included many changes in working practice, the inclusion of stronger consensus management and many interventions aimed at increasing safety 'awareness'.

The management of key processes, the second general category of interventions, is amenable to more traditional quality improvement methodologies including the use of run charts and PDSA improvement cycles. During this programme, we coined the term 'process reliability' to describe the importance of managing the subtasks necessary to ensure a particular outcome. To take a simple example, to reduce the incidence of pressure ulcers in patients, we would need to be sure that the process of initial patient assessment takes place, that correct equipment is provided, the patient turned as required and so on. Failure in each of the key processes can lead to pressure ulcers—usually measured as adverse outcomes.

In managing these programmes at Trust level, and because of the need to avoid a proliferation of interventions, Trusts usually created a 'safety set'—a core set of those factors that reduced risks and—unlike discrete changes in working practices—were addressable through process reliability measures. For example, the safety set applied in a Trust developing safer handovers in a surgical pathway comprised the following six measures, each relating to the risks identified during the diagnostic phase:

1. % patients with senior medical doctor pre-operative review
2. % patients with documented pre-operative surgery review and plan
3. Time between same day cancellations for dialysis patients
4. % theatre IT documents completed with correct surgeon and WHO checklist information
5. % patients with documented post-op day 1 surgical visit
6. Length of stay and readmission within 30 days.

Throughout the course of the programme, the safety set monitored these metrics. Figure 6.11 provides an illustration of item 3:

The design of interventions in improving patient safety can be problematic. Clinical teams are usually unable to simply put the system on hold, while better systems are designed and implemented and must continue to deal with the needs of patients and the flow and pressure within the patient pathway. For this reason, many—or even most—improvement initiatives have tended to be based on familiar cycles of continuous improvement. In classical management of risk, a number of distinct strategies are usually open, which include avoidance, transfer, mitigation and, best of all, elimination. Redesign of systems is ideally a method for removing the risk completely. To that end, we believe that safety interventions made either proactively or reactively should use the principles embodied in the 'hierarchy of control' where possible.

The hierarchy of control provides a framework to inform what interventions can be made for best effect.

- Eliminate risks—substitute hazardous equipment and/or processes with ones that are inherently safer.
- Contain risks—design equipment and processes to protect users from the hazards.

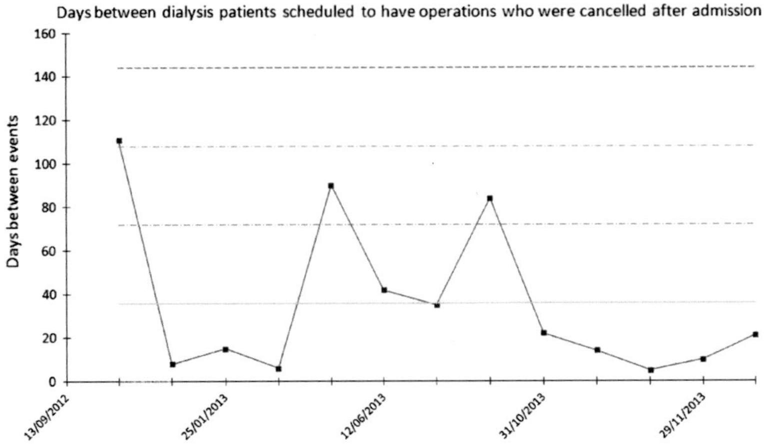

Fig. 6.11 Monitoring process reliability in safety set

- Minimise risks—institute suitable systems of working.

6.15 Uncovering Risk—A Platform for Safety Management

Whatever the actions taken as a result of the diagnostic phase of the programme, and their relative effectiveness, any serious attempt to prevent future harm to patients must be based on a detailed appraisal of risk. Of course, risks are highlighted every day in healthcare systems when the systems fail and an incident or near miss occurs, to the detriment of the patient. But approaching risk management solely on this basis has not proven successful in preventing harm. By using established techniques of analysis, by unlocking the tacit knowledge of multidisciplinary teams and by the open evaluation of error potential in chosen pathways, the teams engaged in this programme took a different approach. As a result of their proactive analysis, their interventions could be directly linked to systems risk and targeted at its management—through redesign, improvement cycles or 'cultural' interventions such as communication, huddles and the raising of the consciousness of safety.

Engaged teams of clinical and governance staff will often be aware of risks in systems as part of their practice, of course. Critically, though, they may not share a consensus view of their magnitude or agree on which risks present the most significant threat to patient safety; without a systematic approach as applied through these methodologies, safety interventions are without a proactive platform and may also be unsighted on hidden risks. Crucially, in the analyses carried out during the Safer Clinical Systems programme, many teams identified previously unrecognised risks. For example, in one Trust working on medication safety, a lack of medical and nursing staff's knowledge of the differences between standard and modified release

Table 6.2 Risk Reducations in Trusts applying proactive methodologies

Site	No. of risks reported	Number with reported reduction
Bristol	6	2
Bath	12	12
Birmingham	3	2
Dumfries	9	3
East Kent	16	7
Manchester	12	10
Nottingham	18	12
Salford	4	2
Total	80	50
Site	**No. of 'high' risks reported**	**Number with reported reduction**
Bristol	6	2
Bath	10	10
Birmingham	2	2
Dumfries	1	0
East Kent	9	6
Manchester	9	8
Nottingham	16	10
Salford	4	2
Total	57	40

formulations was uncovered, and also analysis tools helped to unveil where there are inefficient processes for ordering and supplying medicines which leads to missed and delayed doses. Overall, in five out of eight task analyses, hidden risks were brought to light.

In terms of risk reduction, at the conclusion of the programme, the eight sites who initially took part in this exercise reported a total of 80 systematically identified risks where the existing control measures were weak or absent and where interventions were designed and introduced to reduce risks to patients. Of those 80, the sites reported that 50 had been reduced (Table 6.2).

Looking only at risks categorised as 'high' during the diagnostic phase, and excluding any ambiguous current risk evaluations (where a risk was categorised as high/medium, for example), the sites identified a total of 57 risks where control measures were inadequate. Of these, sites reported a reduction to either 'low' or 'medium' in 40 cases. Of these, 12 key risks were currently evaluated as 'low'.

6.16 Residual Risks—Escalation and Governance

While these methodologies can be used to identify risk and provide a platform for safety, it would be naïve to believe that all risks in a complex, pressured healthcare system can be adequately controlled. In many cases, an honest review of risk concluded that, even after sustained interventions, 'residual risk' remained.

What can frontline staff do when they identify significant residual risks as a result of this work that they feel cannot be changed—the big problems that are just too difficult to address? What if there is no resource to provide extra pharmacists? What if the organisation simply cannot locate and recruit or even afford the nurses needed for the night shifts?

This is where a risk-based approach and the processes of escalation in an organisation assume their real significance. Where risks are identified, the programme focused on the elimination of risk through system redesign, or containment of risks, or minimising their effects (described as the hierarchy of control). However, some risks cannot be easily addressed from the sharp end alone.

Risk awareness and good governance are essential. Uncontrolled risks have to be managed appropriately. In most cases, they will be placed on the risk register of the organisation—sometimes used wisely as a live document and improvement tool, but sometimes a repository of things we can do nothing about but that continue to worry us. We believe that uncontrolled risks to patients should be escalated and reviewed honestly through the various layers of an organisation. In most cases, a senior manager (probably an executive board member) must take the responsibility for how to deal with uncontrolled risks to patients. He or she may decide, for example, to simply transfer the risk to another Trust and cease to offer the services concerned; they may decide that it is better to try and mitigate the risk through detailed resilience planning and day-to-day monitoring; or the organisation may simply decide to accept the risk. In those cases, the escalation of risk must go beyond the healthcare organisation concerned—to those who commission, regulate or fund.

The thread of risk goes from the sharp end of clinical practice all the way through the organisation and its executives, and on to regulators and commissioners. Building Safer Clinical Systems for patients has to acknowledge this risk, do what can be done through the techniques described in this reference guide and escalate those risks that remain—wherever that may lead.

The following section provides some insight into a tool—the safety case—that can serve as a transparent platform for risk and safety management and explicitly addresses risk control in a system and residual risk management.

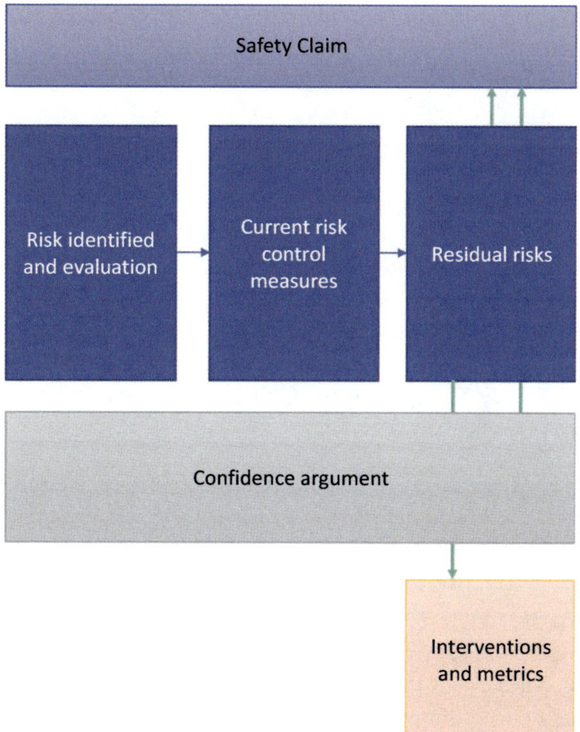

Fig. 6.12 Conceptual diagram of safety case

6.17 The Safety Case

Overview

In building safety in complex systems, we adopted many of the tools and techniques already used in other safety-critical industries, one of which is a formal statement and summary of the safety of a system known as a *safety case*. This tool can serve many purposes. It can be used as a snapshot of safety, showing how safe a team believes the system to be here and now: it can be used as a tool in improving safety; it can be shared to build knowledge and consensus about safety. And it could be used, in a short form, as an assurance statement for internal and external stakeholders.

A conceptual representation of a simple safety case, slightly simplified from that used in supporting Trust teams in evaluating safety, is shown in Fig. 6.12.

The simple form of this example reflects very closely the underlying processes of this proactive approach to safety. Firstly, a safety case makes a *safety claim* relating the level of safety. This is, of course, built from the evaluative processes described earlier through *risk identification and evaluation* in the pathway or area concerned and the *risk control measures* already in place. A natural outcome of this is an understanding of risks which are inadequately controlled—the *residual risks*.

Ideally, those residual risks become the target of *interventions and metrics*, summarised as ongoing measurements of safety, which were dubbed during the programme as the 'safety set'—a collection of process reliability measures, for example, that might relate to the performance in key risks of the pathway. In medicines management, for example, they may include risks in medicines reconciliation, prescribing accuracy and so on.

In this short-form case, we also include what is known as a *confidence argument*. In essence, this represents the degree to which the risk evaluation process is robust—do we really understand the risks in the pathway and on what basis. The confidence argument serves as a check or an internal reflective process to challenge the team's understanding and ensure that key risks to patients are recognised and the outstanding risks are addressed.

6.18 Safety Cases in Practice

At the conclusion of the safety programme, we conducted semi-structured interviews with representatives of each Trust to review the use of proactive safety methodology generally and the use of safety cases specifically. We found that the 'safety claim' can be a difficult issue to determine or even address, since the level of acceptable risk in a clinical system is an area fraught with difficulty. As an example, we might consider events in the NHS which are deemed easily preventable such as wrong-site surgery and other surgical errors, collectively known as surgical 'Never Events' and closely monitored by regulators. The national frequency of these events in the three most recent years for which data are available (2013/14–2015/16) was about 1000 in more than 30 million procedures—a rough probability of 3×10^{-5}. A recent review of this suggests that this level of risk probability is extremely small and that in other systems would be subject to robust scrutiny and challenge by a regulator. Is this an acceptable level of safety in clinical practice and would a safety claim of this nature be accepted as 'safe' in common understanding? One answer is 'probably not', since each of those 1000 events is a patient who may have been harmed (not all never events result in harm; this is a measure of error, not harm), with all the human consequences implied.

Though the safety claim element may be both difficult to arrive at and of uncertain utility, in a clinical system the proactive analysis of risk (and the uncovering of residual risk) is essential in understanding patient safety. All participating sites reported benefits of this approach.

> [The safety case] has been incredibly helpful, and I would say successful as well, I think it's given us clear focus on what the problems are. It's given us a voice at Trust Board level. That's the bit I've been really shocked in how powerful it's been. It's given a tighter vehicle that you can take to frontline staff and up to Board as well. And people understand it. (SCS team leader, Acute Hospital Trust)

Using the safety cases requires safety teams to make explicit links between safety interventions and the risks—sometimes not previously recognised—identified during system diagnosis.

> I'll give you an example of the elective pathway where there was no medication chart written at all prior to patients going to theatre and patients also, the nurses were not asking for patient's medication information at all. Again at the beginning, we did a lot of work around getting the, trying to get patients to bring their medication information to pre-op assessment at the start, so we did a campaign in the media. We did posters in 200 GP surgeries and chemists in the area. We went to the local health visitors groups that recognizes actually everybody needs to bring their medication information in. That helped us enormously. We got all the nurses trained up, we got the administrative staff trained and as soon as patients came in, if they left them, which we'd amended to say bring your medication information in hasn't worked. (Pharmacist, SCS participant)

Interventions used in response to the risks highlighted varied, though all sites reported that they had introduced system or pathway redesigns, with three sites also emphasising the need for better process reliability in key tasks. Sites also highlighted that where redesigns had taken place (such as introducing additional checks based on the risks identified earlier), they still had to build reliable implementation of the tasks.

> I think we've gone more for the increased reliability side rather than being able to design a step out. We've designed a new step that's to identify people at risk, which we kind of hadn't had before, which was the un-clerked patient. So now one of the three questions by the nurse is "Do you take medication for diabetes, Parkinson's Disease or Epilepsy? If so, when is your next dose due? Have you got your meds with you?" (Consultant surgeon)

The establishment of risks and the uncovering of hidden risk are clearly valued. What are the best sources and tools for this process? Information sources used in the process of system diagnosis and in the construction of a summary safety case can be varied. We have described the basic analytical processes of process mapping, FMEA and HTA (which would normally include the assessment of human error in systems), as well as the various evaluations of safety culture. Traditionally, clinical teams would also look to incident reports, root cause analyses of patient safety events and audits of quality. In the process of applying these tools, we sought feedback from teams to try to understand the usefulness of these resources in building safety and also the ease of access to clinical teams—the usability of the resources.

In proactive analysis, most teams used several information sources and tools, Table 6.3.

However, the usefulness of these techniques varied. Table 6.4 gives perceived utility in proactive risk evaluation (respondents were asked to rate each item below as very useful (1) to waste of time (5)).

The hands-on expertise and experience of participating clinicians was strongly favoured over the existing risk management systems (essentially reporting and analysis), but the most useful information here was seen to derive from three key techniques—process mapping, HTA and FMEA. These techniques, however, did require a technical understanding beyond that usually available at Trust level—as might be expected in a formal safety analysis.

Table 6.3 Information sources

Element	No. of sites using element
Incident reports	7
Incident investigations	5
CQC essential standards	1
NICE or best practice guidelines	5
Opinions from individual practitioners	8
Opinions from group work with practitioners	8
Existing process maps	3
New process maps	7
HTA	8
FMEA	8
Risk evaluation and ranking	7
Predictive human error analysis	2

Table 6.4 Value of information sources

Element	Mean utility
Incident reports	3.5
Incident investigations	3.4
CQC essential standards	3
NICE or best practice guidelines	3.4
Opinions from individual practitioners	1.8
Opinions from group work with practitioners	2.13
Existing process maps	2.67
New process maps	1.29
HTA	1.5
FMEA	1.5
Risk evaluation and ranking	1
Predictive human error analysis	2.5

6.19 A Safety Case in Medicines Management

An example of a safety case prepared for medicines management in an acute Trust is provided on the following pages. The analysis was based on an initial review using a process map, followed by a detailed hierarchical task analysis—shown earlier in section x. The high-level analysis was followed by detailed deconstruction and risk evaluation, including a review of performance influencing factors.

For illustrative purposes, we have based the safety case on high-risk tasks only, and these and the other elements of the safety case are given in Table 6.5.

Table 6.5 Medicines management: evidence of risk

ID	Description	Risk	Risk evaluation	Current risk control measures	Residual risks	Interventions required	Key metrics
1	Reconcile patient medications *within 24 h of admission*:						
1.1	Confirm the existing patient medication	Incorrect medicines reconciliation	H	Doctor or pharmacist checks prescriptions with patient, family, GP or care home	Inadequate pharmacy cover leads to incorrect reconciliation	Increased pharmacy cover	Numbers of patients with reconciled medications on admission
1.2	Record details of patient medication on drug history	Transcription errors	H	None	Transcription errors lead to incorrect medication	Consultant check of transcriptions on ward round	Transcription errors reported
1.3	Check that medicines, doses and frequencies are correct	Incorrect medication or dosage	H	Check by pharmacist or doctor	Checking by doctors rarely carried out, leading to incorrect medication or dosage	Increased pharmacy cover	Numbers of patients without daily medication check
2	Prescribe required medication:						
	2.3.5.1 Write drug name clearly	Incorrect medication	H	None	Poor-quality writing of prescription leads to incorrect medication	Consultant check of drug chart on ward round	Number of patients without consultant check

(continued)

Table 6.5 (continued)

ID	Description	Risk	Risk evaluation	Current risk control measures	Residual risks	Interventions required	Key metrics
2.3.7	Add a contact or bleep number	Lack of check leads to incorrect medicine or dosage	H	None	Lack of contact details leads to incorrect medication or dosage	Consultant check of drug chart on ward round	Number of patients without consultant check
6	Prepare for administration of drugs:						
6.2.1	Ask patient name	Medication incorrect for patient	H	Training and supervision of nurse drug round	Training and supervision shortfalls lead to incorrect medication or dosage	Refresher training in drug administration	Percentage of nurses completed refresher training
6.2.2	Ask DOB	Medication incorrect for patient	H	Training and supervision of nurse drug round	Training and supervision shortfalls lead to incorrect medication or dosage		

(continued)

Table 6.5 (continued)

ID	Description	Risk	Risk evaluation	Current risk control measures	Residual risks	Interventions required	Key metrics
6.2.3	Check nameband	Medication incorrect for patient	H	Training and supervision of nurse drug round	Training and supervision shortfalls lead to incorrect medication or dosage	Refresher training in drug administration	Percentage of nurses completed refresher training
6.5.1	Ensure patient's weight is stated	Medication incorrect for patient	H	Training and supervision of nurse drug round	Training and supervision shortfalls lead to incorrect medication or dosage	Refresher training in drug administration	Percentage of nurses completed refresher training
6.5.3	Check relevant tests	Medication is missed or not modified	H	Training and supervision of doctors and nurses Pharmacist check on pharmacy ward rounds	Inadequate performance or resource leads to incorrect medication on administration	Refresher training in drug administration Increase pharmacy cover	Percentage of nurses completed refresher training Number of patients without daily medication check
6.5.4	Determine suitability of medication based on observations and test review	Side effects or efficacy of medication is not observed and acted upon	H	Nurse training Pharmacist checks on pharmacy ward rounds	Lack of nursing skills and resource leads to missed side effects Lack of pharmacy support leads to missed side effects	Refresher training in drug administration Increase pharmacy cover	Percentage of nurses completed refresher training Number of patients without daily medication check

(continued)

Table 6.5 (continued)

ID	Description	Risk	Risk evaluation	Current risk control measures	Residual risks	Interventions required	Key metrics
6.6	Select medication:						
6.6.1	Select from trolley	Selection of wrong medication	H	Red tabards on some wards to protect nurses from distractions Nurse training and supervision	Not all wards use red tabards Variable practices in drug rounds Similarity in drug packaging Poor design and usability of trolleys	Increased use of red tabards Standardise drug rounds in consultation with nurses and pharmacists	Number of wards without red tabards Revised protocol agreed Revised protocol included in training
6.6.7	Establish the dose	Wrong dose selected	H	Red tabards on some wards to protect nurses from distractions Nurse training and supervision	Not all wards use red tabards Variable practices in drug rounds		
7	Administer each medication						
7.2	Observe medication taken	Drugs not administered	H	Red tabards on some wards to protect nurses from distractions Nurse training and supervision	Not all wards use red tabards Variable practices in drug rounds	Increased use of red tabards	Number of wards without red tabards

Medicines Management Pathway
Acute Hospital Trust

Safety Claim

The medicines management pathway has a **moderate** level of safety.

During the year 2014/15, one patient died as a result of medication error and 4 serious incident investigations were required as a result of medication incidents.

Evidence of Risk

Pathway risk was examined through process mapping, hierarchical task analysis and failure mode and effects analysis. Past incidents and investigation were also reviewed.

Residual Risk

Residual risks are shown in the analysis above and expressed qualitatively.

Most significant risks relate to:

- Lack of pharmacy cover leads to a failure to meet Trust standards of daily pharmacist review of all patients.
- Lack of pharmacy cover and junior doctor resource leads to errors and delay in medicines reconciliation.
- Unstructured consultant ward rounds do not include review of drug charts, leading to potential errors in prescriptions and administration of medicines.
- Interruptions of drug rounds by staff and patients lead to errors in the administration of medicines.
- Lack of consistency, training and supervision of drug rounds lead to errors in the administration of medicines.
- Lack of review and observation of patient responses to medication leads to potential suboptimal use of medications.

Interventions and Metrics Required

To address the highest risks in the overarching medication pathway across the Trust, the following interventions are in place and reviewed monthly at Patient Safety Committee:

Systems Changes

1. Increase the use of red tabards for drug rounds to all wards (Chief Nurse).
2. Include medicines review in all consultant ward rounds (Medical Director).
3. Review drug round protocol (Deputy Director of Nursing).
4. Seek funding to increase pharmacy cover (Board of Directors).

Process Reliability Metrics

1. Numbers of patients without medicines reconciliation within 24 h of admission
2. Percentage of nurses trained in revised protocols for drug rounds
3. Percentage patients receiving daily consultant review.

Confidence Statement

This analysis has been developed based on a review of the existing policy documents, incident investigations relating to medication error, a series of group discussions with stakeholders (ward leaders, pharmacy technicians, medical trainees and Band 5 nursing staff) and workplace observations on three wards. Observations and HTA were conducted by external human factors experts and validated with the trust safety and quality and risk management teams. We have a high degree of confidence in these findings.

However, while all wards are technically subject to the same medication management policies, there are variations in practice that were discussed during the analysis. For some critical subtasks, it may be beneficial to develop more detailed task analyses to identify variations in practice between wards, and consider how these variations may affect risk. The analysis assumes a standard progression through the medication management process. There may be situations where standard progress does not happen, and this in turn may have an impact on risk.

6.20 Safety Cases and Regulation

Where patient safety is under scrutiny—often as a result of egregious incidents or, increasingly, higher than expected mortality, there is a legitimate need for external assurance. This is always carried out through direct inspection of providers, based on a review of data and policies. The inspection model has been subject to criticism, some of it justifiable, since it cannot be shown to have reliably improved safety or quality.

Based on personal experience, the goals of inspection are laudable and the depth of inspection is often formidable. It is rare, however, that questions are asked relating to the reliability of key processes (such as patient observations) or about the risks embedded in the systems. The inspected area is not usually required to demonstrate any level of knowledge of system design, the risks relating to human factors, the level of residual risk or the result of escalation processes. This frequently means that regulatory inspection into safety is ad hoc and heavily based on point observations. Of course, where a system has not been internally reviewed as described here as 'system diagnosis', the provider would not be able to comply with a more mature approach to risk and the regulator would have little choice.

But they should. It is a major failing of risk and safety systems that they are not required—as they are in other safety-critical industries—to demonstrate to regulators that they have real knowledge of risk, gathered systematically and owned by the subjects of inspection. The use of a safety case addresses this issue directly, and its lack reveals a loose approach to safety that is based only on passive management. A reactive approach to assurance fails to recognise that a given system may appear safe merely through happenstance and that patients may be harmed in the future, where proactive risk management does not take place. In other words, even a deep dive by inspectors, including a review of past harm, is fraught with potential failures.

Would regulation based on credible, systematic self-review achieve more? It is hard to argue that it would achieve less. And the difficult job of the regulator would have a basis in more than observation; it would be based on a challenge to the existing safety analysis, which is a different matter entirely.

6.21 Concluding Remarks

This chapter has described how the Safer Clinical Systems programme has been applied in some NHS acute Trusts. We have aimed to show that, despite a strong intention for organisational learning—a process which began in the NHS more than a decade and a half ago and the subject of recent re-emphasis—the reality of the existing systems is deeply problematic. These reactive systems of risk management and improvement take place in a dispiriting culture of blame and bureaucracy, with scant knowledge of human factors, human error or system risk.

And yet, surveys show a strong commitment to improving safety on the part of individual staff and a clear recognition of how systems create the ground for patient harm and human error. In the area of organisational learning, the system not only does not work but also actively frustrates the good intentions of clinicians and managers bravely trying to make things better.

In this environment, the proactive approach offers a way of creating safety before patients are harmed—and it does so in a way in which blame is completely irrelevant. It uses tools which break down, rather than reinforce, barriers between professions—tools which rely on consensus, which surface tacit knowledge and create a shared perspective of risk and safety.

Some of these tools require a deeper understanding of systems and analysis than that required to react to harm through patching up a broken system. None of them rise to the level of inaccessibility, however, and all are routinely used in other safety-critical industries. The proactive approach takes a benign view of human behaviour; it seeks to identify performance influencing factors, to remove the traps in the system that create error and recognises that humans can also create safety in their work.

We have seen that the proactive approach, moving through an assessment of culture, a phase of system diagnosis and a continuous phase of systematic improvement, not only reduces the prevalence of risk to patients, but uncovers hidden risks. Participants in this programme experienced a growing confidence in the knowledge of risk and safety. While they certainly did not believe that 'everything was ok', they knew more than when they began and they had the tools to understand still more. A reliable organisation, a learning organisation, *needs* these qualities: a focus on front-line knowledge, a respect for all professions and a searching-out of vulnerabilities and potential problems.

Not all problems can be solved. In the example safety case we presented here, for example, the level of pharmacy cover (with its consequences for budgets and the difficulties in recruitment) was probably impossible to address from within the local team. In most systems, this would be raised with senior management and possibly placed onto the risk register, where it would perhaps languish with hundreds of other risks until funding was released or patient harm forced the hand of a heavily constrained Board of Directors. This is why we have emphasised the use of 'residual risks' as a concept and as a component in a safety case—accompanied by the use of a formal escalation process. Risks to patients cannot go unrecorded but must be clearly stated, ranked and based on credible evidence to receive the attention they warrant, at whatever level that might be, within the department, directorate, Trust or beyond.

The lessons of national investigations into patient harm and local investigations into safety incidents have yielded necessary interventions and useful debate. There is a sense, however, that the service is in danger of accepting that we will always be fire-fighting, beleaguered by the need to react and never able to really plan for safety. This programme has, for the first time in the safety work of the NHS, attempted to move the approach from one of wait-and-fix to one where we are able to manage our risks in advance, recognise what risks are unaddressed and provide credible arguments to commissioners, regulators and government of active safety management.

References

Franklin, B. D., Shebl, N. A., & Barber, N. (2012). Failure mode and effects analysis: Too little for too much? *BMJ Qual Saf*, 21, 607–611.

Failure Modes and Effects Analysis (FMEA) Tool; Institute for Healthcare Improvement; http://www.ihi.org/resources/Pages/Tools/FailureModesandEffectsAnalysisTool.aspx

The Manchester Patient Safety Framework (MaPSaF). / Parker, D., Kirk, S., Claridge, T., Lawrie, M., & Ashcroft, D. M. (2007). Patient safety research: Shaping the European agenda—International Conference Porto, Portugal. Available at: https://www.research.manchester.ac.uk/portal/en/publications/the-manchester-patient-safety-framework-mapsaf(e8c48571-111a-4eac-afd0-a7bd361ee112)/export.html#export

The Hierarchy of Control, https://www.cdc.gov/niosh/topics/hierarchy/default.html

Safe Today–Safer Tomorrow, Cross, S., & Ball, Z. NHS Quality Improvement Scotland. http://www.healthcareimprovementscotland.org/his/idoc.ashx?docid=7a9246be-3a6f-4085-b02b-760e467fbeea&version=-1.

Safer Clinical Systems, a reference guide for clinicians and managers. (2013). http://patientsafety.health.org.uk/sites/default/files/resources/hf_safer_clinical_systems_reference_guide_final_1.pdf

Spurgeon, P., Flanagan, H., Cooke, M., Sujan, M., Cross, S., & Jarvis, R. (2017). Creating safer health systems: Lessons from other sectors and an account of an application in the Safer Clinical Systems Programme. *Health Services Management Research*, 1–9.

Stanton, N. (2006). Hierarchical task analysis: Developments, applications, and extensions. *Applied Ergonomics*, 37(1).

Sujan, M., Spurgeon, P., Cooke, M., Weale, A., Debenham, P., & Cross, S. (2015). The development of safety cases for healthcare services: Practical Experiences, opportunities and challenges. *Reliability Engineering & System Safety*, 140, 200–207.

Westrum, R. (1984). *Complex organizations: Growth, struggle, and change*. Englewood Cliffs, NJ: Prentice-Hall, Inc.

Westrum. (2007). NHS Scotland Independent and Reporting Culture. www.nhshealthquality.org.

Chapter 7
A Practical Effective Tool for Measuring Patient Safety Culture

7.1 Introduction

As suggested in Chap. 4 the culture in an organisation can be a crucial part of how patient safety is tackled and in order to pursue this in the implementation of Safer Clinical Systems, a specific measure (Safety Culture Index) was used. This chapter outlines the development of this measure and the results of its implementation in the eight sites. Safety culture, even if considered as part of a wider organisational culture, is a yielding and often disputed term. Nonetheless, there is increasing advocacy of building a positive safety culture as a critical component of hospitals becoming high reliability organisations (Singla et al. 2006; Pronovost et al. 2006). Environments outside of health may well face hazards and risk on a daily basis but manage to sustain their levels of safety by building an appropriate safety culture (Leonard and Frankel 2012). The role of safety culture is perhaps captured most impressively and succinctly by Leape and Berwick (2005) who say

> the combination of complicity, professional fragmentation, and a condition tradition of individualism enhanced by our well entrenched hierarchical authority structure and diffuse accountability, forms a daunting barrier to creating the habits and beliefs of common purpose, teamwork and individual accountability for the successful interdependence safety culture requires. (p. 2385)

The complex interplay between individual and group values, attitudes and patterns of behaviour not only determine organisational commitment to safety issues but also shape the appropriateness of safety management strategies. The crucial role of a positive safety culture lies in its potential motivational impact on healthcare professionals to adopt appropriate safety attitudes and select behaviours that enhance patient safety. In the context of the implementation of the SCS model across the eight NHS sites (see Chap. 5 for an account of the SCS process), it was felt to be important that there should be an assessment of the safety culture existing in the organisations, and indeed in the specific departments or patient populations that were to be the focus

P. Spurgeon et al., *Building Safer Healthcare Systems*,
https://doi.org/10.1007/978-3-030-18244-1_7

of the programme. If the attitudes, beliefs and values of the staff groups in each site may act as a mediator to potential safety initiatives, then it is almost self-evident that such knowledge about them may enhance the likelihood of successful introduction and implementation. There was therefore a need to assess the existing safety cultures in the collaborating sites.

7.2 Measuring Patient Safety Culture

An early consideration was the Manchester Patient Safety Framework (MaPSaF) developed and published by the University of Manchester (2006). The tool was developed from a series of interviews and focus groups and usefully describes in organisational headlines the nature of the patient safety culture in the organisation. Details of the five states/levels within MaPSaF are described in Chap. 6.

This is a useful structure and the SCS team did use it (see Chap. 6) and largely the sites could be labelled Reactive to Bureaucratic. This information was derived from small subgroups of staff from each site. However, as the originators of the tools suggest it is primarily aimed at provoking discussion and while this can be very valuable, it did not provide the measurement of safety culture that was being sought.

In pursuit of a more quantitative measure, the team was aided by a number of very thorough reviews of approaches and tools used to measure patient safety culture (Singla et al. 2006; Flynn et al. 2006; Health Foundation 2011). Virtually, all surveys assess core dimensions of safety such as communications, teamwork and leadership for safety. However, surveys differ in the degree to which they address a particular clinical area and the amount of coverage even to specific aspects of safety culture. Although safety leadership is almost an ever present, it is worth recognising that leadership of doctors in healthcare settings may be more problematic than in other sectors (Flynn et al. op.cit; Spurgeon and Clark 2018).

The report from the Health Foundation (2011 is probably the most comprehensive review as well as being the most recent. The summary in this report is most salient. They found that many of the tools lacked the psychometric properties needed to for improvement assessment. There was limited evidence of outcome validation with one or two exceptions such as the studies by Robb and Seddon (2010) and Cigularov et al. (2010). The majority of the measures were from the USA with norms also either from the USA or from across-country amalgamations which can obscure their relevance to a potential country or setting. This review was particularly aimed at helping practitioners select the survey for use, and their concluding statement is telling:

> No single tools stood out as being the most useful for organisations in the UK. (page 35)

The description of the components of culture provided by the Institute of Healthcare Improvement (Frankel et al. op.cit.) is very useful but the attempt here is to turn these high-level concepts into measurable information.

7.3 Developing the Safety Culture Index (SCI)

In the light of the conclusion that there was no single 'best' tool for assessing safety culture, it was decided to build upon previous work involving one of the current authors (Spurgeon et al. 1999) and develop a bespoke measure for the project.

Shared group perceptions define the culture of the organisation and have the potential to guide how members of staff think, act and feel. Although the measures of safety culture focus on attitudes and are sometimes described as measuring climate rather than culture, there are numerous studies linking responses to such measures to safety-related outcomes. The terms safety culture and safety climate are generally used interchangeably. Culture may be thought of as relatively deep seated with climate a more surface representation of it. Importantly though climates seem more responsive to intervention and as a consequence affects culture.

It is important to recognise that in complex large organisations, there is unlikely to be one universal good culture. Since culture reflects the attitudes, beliefs and perceptions of individuals in the organisation, it is clear that these may be shaped by the delivery context. An emergency department is different from a rehabilitation ward which in itself is unlike an intensive care unit, and so on. It is well established that a large organisation may have many subcultures. What is important is to identify an appropriate safety culture that operates to ensure safe delivery in the particular work setting (Flynn et al. 2006). There are some important factors which influence workplace safety, and these can operate at different levels.

Individual influences

- Staff satisfaction and morale
- Frequency/quality of performance appraisal
- Concern for standards and pride in work
- Keenness to learn from errors or adverse incidents
- Degree of self-esteem and perceived empowerment
- Autonomy and participation in decision-making

An individual's profile across these issues will influence how they see their personal responses to safety-critical behaviours.

Interpersonal influences

Opportunities to openly discuss issues
Group norms condoning safety violations
Reality of blame-free climate
Extent of collaboration between staff groups
Conflict between management and other staff

The ethos surrounding a team in relation to these factors will contribute to a more collective view of safety behaviours.

Organisational influences

Clarity of role and responsibility expectations
Quality of supervision and standards monitoring
Availability of appropriate equipment
Level and quality of training

An organisations 'profile on these will represent how far the leadership of the organisation has succeeded in creating a priority for safe care delivery.

7.4 Scope for Service Improvement

Degree of Bureaucracy

The SCI brings these ideas about the safety culture assessment together by operating at an individual, team and organisational level. Its design also closely reflects and reinforces the thinking that underpins Safer Clinical Systems by encompassing the three core concepts of

- Systems thinking
- Proactive search for risk
- Feedback to support organisational learning.

The SCI comprises a matrix which when scores are entered provides a safety culture profile which can be analysed by staff group or by different areas of organisational activities. The matrix is presented (below) and is formed by the individual, team and organisational levels and then by task, people control and change orientation. Each of the scales within the 12 scale matrix is defined briefly. A recent research scan of safety culture measures by The Health Foundation (2011) noted that many of the measures available lacked evidence of reliability or validity. The SCI in contrast has scale reliability (lowest reliability coefficient of 0.7) and in previous work has been shown to link to incidents in NHS Trusts (Spurgeon et al. 1999).

SCI© Scale Definitions

	TASK & TEAMWORK		MAINTENANCE & IMPROVEMENT	
	Task Focus	People Focus	Control Focus	Change Focus
INDIVIDUAL *('Maintain Safety Competence')*	*Scale a1)* **Coping with work demands** This scale is concerned with the extent to which a member of staff feels capable of working safely when coping with variations in workload and increased levels of job stress.	*Scale b1)* **Participation in decision-making** This scale assesses the extent to which the individual is able to act independently and take responsibility for dealing with problems and making decisions about salient issues at work.	*Scale c1)* **Checking & accountability** This scale is concerned with the extent to which individual members of staff actively take responsibility for making sure they monitor and maintain safe working practices.	*Scale d1)* **Commitment to learning** This scale focuses upon the individual's effort to learn from errors and mistakes made in the workplace and their desire to improve their level of safe performance.
TEAM *('Enhance Safety & Productivity')*	*Scale a2)* **Purpose & direction** This scale is concerned with the working practices related to the safety rules, policies and procedures which are shared by members of the work group or team.	*Scale b2)* **Working in collaboration** This scale is concerned with the extent to which interpersonal factors impact upon effective teamwork and safe collaborative working in groups.	*Scale c2)* **Sharing information** This scale centres upon the importance to safety of the accurate sharing of work-related information and instructions within the work team.	*Scale d2)* **Blame-free climate** This scale is concerned with the psychological safety and security within the work team aimed at protecting the individual from a destructive level of blame.
ORGANISATIONAL *('Provide Safety Leadership')*	*Scale a3)* **Role clarity** This scale focuses on the perceived extent to which the organisation encourages and reinforces staff to perform effectively with respect to matters of safety.	*Scale b3)* **Staff motivation** This scale relates to the extent to which management are friendly and supportive to staff and actively motivate them to adopt safe working procedures and practices.	*Scale c3)* **Standards monitoring** This scale assesses the extent to which the organisation informally and formally monitors staff performance and provides constructive feedback about safe working practice.	*Scale d3)* **Vision & mission** This scale centres upon the extent to which the organisation takes managing safety seriously and makes real efforts to promote the centrality of a safe working practice ethos.

Each profile or cultural type may be relatively safe or unsafe the point depending on the context and variables at play. In general terms, the SCI provides insight as to where safety problems may originate. In the context of Safer Clinical Systems, SCI serves two particular functions.

(1) By defining staff and subsite safety culture profiles, the participating sites had some insight as to how interventions identified by the Safer Clinical Systems process might be received and implemented in a particular safety culture context.
(2) To act as a before and after measure of culture and thereby offer some evidence as to whether the conduct and implementation of the Safer Clinical Systems process might have had some impact upon safety culture in the area of focus to the participating sites.

Finally, we will consider SCI data in action. It is not appropriate in this general paper to provide the actual data of the Safer Clinical Systems participating sites. However, the results of using SCI can be simply represented as High, Medium and Low with the percentage of respondents falling into each category. For example, at the organisational level, the scale '**Vision and Mission**' identifies how well staff believe the leadership of the organisation has promoted and established the goal of a safer service. In most Safer Clinical System sites, about 30–40% of staff felt that this was being done well. However, in another side over 70% felt that this was done really poorly. There is a clear message here to the organisation's executive team with responses of this type. Such data could be from staff groups of different specialties. In other sites, there were marked differences particularly at an individual level of how certain staff groups saw their responsibilities in terms of the safety of service delivery. The value of such information to prompt intervention before failures manifest is self-evident.

In addition, the SCI provides an overall safety index within a range of 1–100. A score of 45 or below represents significant concern and over 55 indicate a more positive safety culture. The profile in the 12 scale matrix therefore provides a very easily accessible indicator of safety culture in particular staff groups all settings. The Safety Index score is adjusted as normative data for the SCI are accumulated.

The SCI is a quick easy-to-use tool that assesses organisational safety culture at a specific and meaningful level.

An illustration of how SCI may be used in the future in any health organisation is included here with simple displays of how it may quickly guide and influence organisations to improve their safety culture. This advisory Framework for use of SCI are set out in the associated manual "Guidelines on using the Safety Culture Index" (Applied Research 2011)

(a) Purpose of these Guidelines

The aim is to describe how to interpret the output of the Safety Culture Index (SCI) © in order to:

- ***Evaluate the level and extent of safety culture***
- ***Explore the implications of diagnosing safety issues***
- ***Inform the selection of organisational intervention strategies.***

A number of key cultural factors relating to poor organisational performance have been identified and the most important of these factors have been incorporated into the design of the Safety Culture Index (SCI) ©. This instrument is based on a psychometrically rigorous framework developed as a broad benchmarking tool to analyse both weak and strong aspects of safety culture in healthcare organisations not as an isolated process but taking into account that organisations are inherently hierarchical and safety culture operates at the individual, the work group level and at the organisational levels.

(b) Structure of the Safety Culture Index (SCI) ©

The current version of the Safety Culture Index (SCI) © consists of 60 items that form 12 scales (shown in the figure below) and provides a multidimensional profile of organisational safety culture. The SCI © scales have been shown to be both reliable and valid and describe the safety-relevant aspects of perceived working practices. The structure of the SCI © provides a conceptual framework that allows everyone in the system to think about and discuss patient risks and hazards and through this learning process becoming sufficiently empowered to help reduce threats to patient safety and actively seek improvements in the quality of patient care.

The SCI © recognises that organisations are inherently hierarchical in structure and the figure below shows safety culture can be assessed at the individual level, the team level and at the organisational level. As well as these three levels, the structure of the SCI also enables safety culture to be assessed within four working contexts. These contexts are called are the task focus, the people focus, the control focus and the change focus.

	Task Focus	People Focus	Control Focus	Change Focus
INDIVIDUAL *('Maintain Safety Competence')*	*Scale a1)* Coping with work demands	*Scale b1)* Participation in decision-making	*Scale c1)* Checking & accountability	*Scale d1)* Commitment to learning
TEAM *('Enhance Safety & Productivity')*	*Scale a2)* Purpose & direction	*Scale b2)* Working in collaboration	*Scale c2)* Sharing information	*Scale d2)* Blame-free climate
ORGANISATIONAL *('Provide Safety Leadership')*	*Scale a3)* Role clarity	*Scale b3)* Staff motivation	*Scale c3)* Standards monitoring	*Scale d3)* Vision & mission

Each of the twelve scales which comprise the SCI are briefly defined in the **Appendix**.

(c) Understanding SCI Matrix Scores

As well as containing scale definitions and sample data detailing the number of returns for each of the surveyed staff groups, the SCI report always contains SCI © Matrix Results for all staff groups who have been survey respondents. An example is shown overleaf to illustrate how the results are fed back for each staff group in turn.

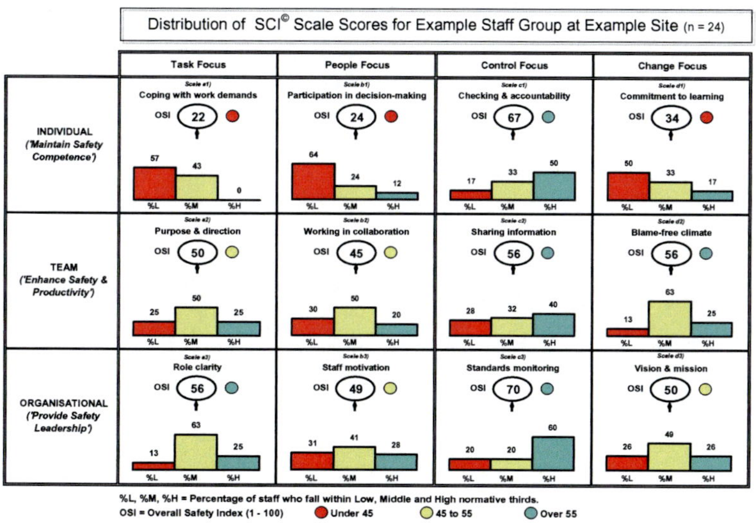

In examining this figure, the following points should be noted:

- The histograms for each of the 12 SCI © scales provide the relative proportion of members of staff whose response were within the bottom (L), middle (M) and top (H) thirds of the SCI © normative database (for a particular scale). Consequently, these summary results for each of the 12 scales are all externally benchmarked and provide a rapid means of examining both the levels and spread of safety culture perceptions. Currently, the norms for the SCI © comprise 1523 healthcare staff in total.
- The single Overall Safety Index (**OSI**) for each scale is not an average score but is calculated from the relative percentage frequencies of respondents who fell into the three normative bandwidths. This figure is then scaled to provide a single summary score ranging from 0 to 100. In this way, the OSI summarises the staff group deviation from the normative mid-point (represented by an OSI score of 50). As a rough guide OSI scores of more than 55 may indicate good safety practices, while OSI scores of less than 45 may signal a need for improvement. An example is shown below:

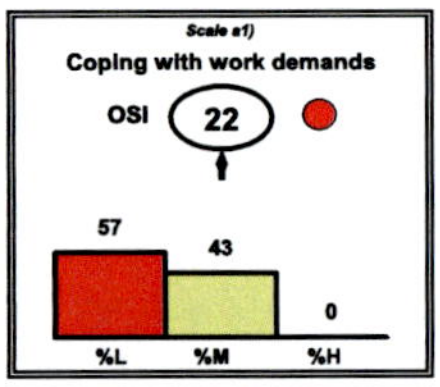

- In this example, it is apparent that 57% respondents rated scale ***(a1) Coping with work demands*** within the low range of the normative database and 43% rated this scale within the middle third of this external norm. It appears that more than half of this staff sample were concerned that coping with work demands might inhibit safe performance of their duties at work.
- The OSI summary score for this scale was 22 (well below the ‘weak’ boundary score of 45) reflecting the fact that most members of staff rated within the worst third of the norms and no members of staff rated within the best third of the norms. This score may signal the need for further investigation to identify problem sources and potential improvement strategies.
- For ease of interpretation, the OSI score has been colour-coded and represented by a dot where:

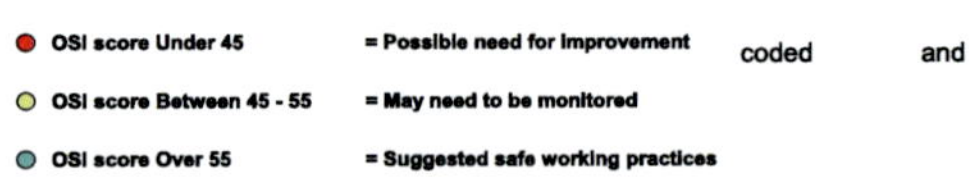

(d) Interpreting SCI © Results

For simplicity of explanation the same example SCI © matrix shown previously on page 114 is shown again but this time without the details of the numeric and graphical data. This simplified version is included below to illustrate the three-step procedure used to analyse the SCI © results.

	Task Focus	People Focus	Control Focus	Change Focus
INDIVIDUAL ('Maintain Safety Competence')	*Scale a1)* Coping with work demands	*Scale b1)* Participation in decision-making	*Scale c1)* Checking & accountability	*Scale d1)* Commitment to learning
TEAM ('Enhance Safety & Productivity')	*Scale a2)* Purpose & direction	*Scale b2)* Working in collaboration	*Scale c2)* Sharing information	*Scale d2)* Blame-free climate
ORGANISATIONAL ('Provide Safety Leadership')	*Scale a3)* Role clarity	*Scale b3)* Staff motivation	*Scale c3)* Standards monitoring	*Scale d3)* Vision & mission

- OSI score Under 45 = Possible need for improvement
- OSI score Between 45 - 55 = May need to be monitored
- OSI score Over 55 = Suggested safe working practices

- **STEP 1: Identify those SCALES associated with positive or negative staff perceptions of safety culture**.

The SCI © comprises twelve scales in all, and the identification of those scales is indicative of areas of both 'strong' and 'weak' safety culture. In this example, it can be seen that five SCI © scales are associated with safe working practices, four scales may require monitoring, and three scales may indicate areas where organisational interventions may be needed.

- **STEP 2: Identify those LEVELS associated with positive or negative staff perceptions of safety culture**.

The SCI © is constructed around three levels (i.e. individual, team and organisational) with each level comprising four component scales. Examining if and where patterns of 'strong' and 'weak' safety culture scales map onto these three levels provides a useful overview of the quality of safe working practices. In this example, it is apparent that all of the 'weak' scales (marked red) occur within the **INDIVIDUAL** level of the matrix suggesting that safety shortfalls may be located at the individual level rather than arising from team factors or poor safety leadership. However, there are four scales at the **TEAM** and **ORGANISATIONAL** levels which may need to be monitored (i.e. they are marked yellow in the table above).

- **STEP 3: Identify those CONTEXTS associated with positive or negative staff perceptions of safety culture**.

The SCI © is based on four contexts where work is focused (i.e. task, people, control and change) with each context comprising three component scales. Briefly, these four **CONTEXTS** are: **TASK FOCUS** (concerned with task achievement, role clarity & clear objectives), **PEOPLE FOCUS** (concerned with participation, staff motivation and information sharing), **CONTROL FOCUS** (concerned with accountability, checking and standards monitoring) and **CHANGE FOCUS** (concerned with commitment to learning and encouraging change). In this current example, all three scales within the **CONTROL** context are marked green indicating that this staff group associated safe working practices with activities such as checking, monitoring, ensuring accountability and maintaining rules and standards.

(e) Characteristics of 'Weak' Safety Cultures

The SCI © is concerned with assessing the shared attitudes, values and beliefs that support both safe and unsafe working practices. There is no one universally 'strong' or 'weak' safety culture and in reality organisational culture reflects many aspects of working practices some of which are potentially unsafe and some of which deserve active support and encouragement through effective management and leadership. Each scale of the SCI © reflects and reinforces numerous organisational characteristics as illustrated in the table overleaf. This table gives some examples of workplace characteristics often associated with 'weak' aspects of safety culture ('strong' aspects are the opposites of these). These examples are not intended to be definitive and should be used to help pinpoint areas of working practices with potential implications for safety issues that may require further investigation.

SCI© Matrix - Characteristics of 'Weak' Safety Culture

	Task Focus	People Focus	Control Focus	Change Focus
INDIVIDUAL *('Maintain Safety Competence')*	Scale a1) **Coping with work demands**	Scale b1) **Participation in decision-making**	Scale c1) **Checking & accountability**	Scale d1) **Commitment to learning**
	'Getting things done' takes precedence over safety concerns	Little interest in discussing safety issues face-to-face with management	Lack of knowledge/awareness about risks and hazards inherent in own job	Shallow reflection and learning from personal mistakes and errors
	Routinely avoiding or failing to fully implement safety procedures	Not keen to participate in decision-making or improving patient safety	Insufficient attention to on-going monitoring of safety of work methods	Little or no effort to maintain/update safety knowledge and competence
	Risky 'cutting-of-corners' due to work pressure and/or poor task planning	Reluctant to provide opinions or give input about safety issues	Not concerned to follow-up outcome of reported safety incidents	Individuals do not actively seek further guidance or training in needed areas
TEAM *('Enhance Safety & Productivity')*	Scale a2) **Purpose & direction**	Scale b2) **Working in collaboration**	Scale c2) **Sharing information**	Scale d2) **Blame-free climate**
	Low shared team purpose and little common understanding of team safety	Limited exploration of work practices to seek ways of improving team safety	Few informal safety discussions and poor attendance at safety briefings	Lack of openness about discussing errors and disciplinary repurcussions
	Inconsistent objectives challenge safety of joint working practices	Low team engagement in making or supporting decisions about safety	Unrealistically low levels of incident reporting rates from team members	Fear of challenging others' risky decisions or unsafe work methods
	Staff provided with little or no clear information on relevant safety matters	Group membership used as 'cover' to avoid taking personal responsibility	Relevant safety standards or updates not communicated or explained	Faults are apportioned or insinuated prior to any incident investigation
ORGANISATIONAL *('Provide Safety Leadership')*	Scale a3) **Role clarity**	Scale b3) **Staff motivation**	Scale c3) **Standards monitoring**	Scale d3) **Vision & mission**
	Staff roles and responsibilities for safety vague or not explicitly stated	Inadequate provision for promoting accurate and timely 'hazard-spotting'	Managers rarely undertake informal 'safety compliance tours' in workplace	Safety issues neither treated as high priority nor given sufficient attention
	Safety training needs analysis not undertaken for all members of staff	Organisational response to reported safety breaches is late or patchy	The incident reporting procedure is too difficult or insecure to be useful	Incidents and investigations are not used to shape effective interventions
	Leadership gives mixed messages re: priority of safety over productivity	Few incentives to empower staff to 'make a real difference' to safety	Organisation safety performance not regularly reviewed at senior levels	Failure to foster environment where safety issues can be openly discussed

The SCI was administered in each SCS site prior to implementation of the intervention and also at the end of the project. The initial administration had two purposes—clearly to establish a baseline for patient safety culture such that any such changes could be assessed using a common measure. It was also intended to guide organisations in approaching the implementation of specific interventions on particular wards or with individual staff groups. The measure provides an insight into differences that may exist between groups even if working in the same area. One might as a consequence anticipate different responses to safety-oriented changes to work patterns or tasks. The SCI profile provides therefore helpful guidance as to the likely receptivity of individuals, groups and teams to changes in work demands and hence enables organisations to plan their these interventions more effectively.

Since the initial use on the SCS project other organisations of gone on to use the SCI as part of understanding the strengths and weaknesses of their organisational safety culture, and this in turn has enabled the development of an increasingly normative database on patient safety to be developed, facilitating meaningful comparison for other organisations.

The findings from use of the SCI in the SCS programme are presented below:

The SCS data were collected at the outset of the programme and also following implementation of the initiatives identified in the SCS process. Six out of the eight sites showed significant positive changes (see table below).

SCI scales	Percentage improvement (%)
Participation in decisions	21
Sharing information	20
Vision and mission	20
Blame-free climate	18
Working in collaboration	17
Checking and accountability	16
Coping with work demands	11

In the context of the Francis Report and other exhortations for the NHS to change its culture, this is important evidence that it can be done. As the colloquialism about culture has it—'culture is how we do things round here' and it would seem the way to change culture in terms of safety is to change the way we do things round here!

Brief Definition of the Safety Culture Index (SCI) Scales

TASK FOCUS

(a1) **Coping with Work Demands**:
This scale is concerned with the extent to which members of staff consider themselves capable of working safely while adapting to variations in workload and coping with increased levels of job stress.

(a2) **Purpose and Direction**:
This scale is concerned with the working practices related to the safety rules, policies and procedures which are shared by members of the work group or team.

(a3) **Role Clarity**:
This scale focuses on the perceived extent to which the organisation guides and encourages members of staff to maintain their expectations of how to perform effectively with respect to matters of safety.

PEOPLE FOCUS

(b1) **Participation in Decision-Making**:
This scale assesses the extent to which the individual is able to act independently and take responsibility for dealing with ongoing problems and making decisions about salient issues at work.

(b2) **Working in Collaboration**:
This scale is concerned with the extent to which interpersonal factors impact upon effective teamwork and safe collaborative working in groups.

(b3) **Staff Motivation**:
This scale relates to the extent to which management is friendly and supportive to staff and actively motivate them to adopt safe working procedures and practices.

CONTROL FOCUS

(c1) **Checking and Accountability**:
This scale is concerned with the extent to which the individual members of staff actively take responsibility for making sure that they monitor and maintain safe working practices.

(c2) **Sharing Information**:
This scale centres upon the importance to safety of the accurate sharing of work-related information and instructions within the work team.

(c3) **Standards Monitoring**:
This scale assesses the extent to which the organisation informally and formally monitors staff performance and provides constructive feedback about their contribution to working practice.

CHANGE FOCUS

(d1) **Commitment to Learning**:
This scale focuses upon the individual's effort to learn from errors and mistakes that they make in the workplace and their desire to improve their level of safe performance.

(d2) **Blame-Free Climate**:
This scale is concerned with the psychological safety and security within the work team aimed at protecting the individual from a destructive level of blame.

(d3) **Vision and Mission**:
This scale centres upon the extent to which the organisation takes managing safety seriously and makes real efforts to recognise and promote the centrality of a safe working practice ethos.

References

Applied Research Limited. (2011). *Guidelines for using the safety culture index (SCI)*. Applied Research Ltd.

Cigularov, K. P., Chen, P. Y., & Rosecrance, J. (2010). The effects of error management climate and safety communication on safety: A multi-level study. *Accident Analysis and Prevention, 42*(5), 1498–1506.

Flynn, R., Burns, C., Mearns, K., Yule, S., & Robertson, E. M. (2006). Measuring safety climate in health care. *Qual Saf Health Care, 5*(2), 109–115.

Leape, L. L., & Berwick, D. M. (2005). Five years after To Err Is Human: What have we learned?. *Jama, 293*(19), 2384-2390.

Leonard, M., & Frankel, A. (2012). Patient safety: How can you ensure everyone plays it safe?. *The Health service journal, 122*(6323), 26–27.

Pronovost, P. J., Berenholtz, S. M., Goeschel, C. A., Needham, D. M., Sexton, J. B., Thompson, D. A., et al. (2006). Creating high reliability in health care organizations. *Health services research, 41*(4p2), 1599–1617.

Robb, G., & Seddon, M. (2010). Measuring the safety culture in a hospital setting: A concept whose time has come? *New Zealand Medical Journal, 123*(1314), 68–78.

Singla, A. K., Kitch, B. T., Weissman, J. S., & Campbell, E. G. (2006). Assessing patient safety culture: A review and synthesis of the measurement tools. *Journal of Patient Safety, 2*(3), 105–115.

Spurgeon, P., Barwell, F., Parker, L., & Dineen, M. (1999). *Organisational Culture and its potential relationship to clinical risk*. HSMC: University of Birmingham.

The Health Foundation. (2011). *Measuring safety culture: The health foundation*. London.

University of Manchester. (2006). *Manchester Patient Safety Framework (MaPSaF)*. Manchester, UK.

Chapter 8
A Systems Approach to Improving Clinical Handover in Emergency Care

Improving Clinical Handover by Understanding How Clincial Systems, Organisational Processes and the Institutional Context Contribute to Vulnerabilities

8.1 Introduction

This chapter describes how the systems approach developed in the previous chapters can be applied to improve clinical handover in emergency care. Communication failures and inadequate handover have been recognised as a significant threat to patient safety for over decade (Johnson and Arora 2009). The Institute of Medicine suggested that poor handover practices were one of the leading causes of medical error in the emergency department (ED) (Institute of Medicine 2007). What is it about handover that makes it such a difficult problem to resolve?

Arguably, one of the key reasons why we have not yet found the right strategies to improve handover practices in a sustainable way is that we have underestimated the complexity of this activity. Handover appears a relatively straightforward task, where one communication partner (the sender) provides the other (the receiver) with certain information. In actual practice, however, handover is much more of a social and dynamic activity than this simple representation suggests. Handover is, of course, about transmitting information—but it is also about much more: it can involve elements of bargaining and negotiation (Nugus et al. 2017), the leadership and the direction of the communication can change to and fro (Sujan et al. 2015a), and it can serve additional purposes such as relationship building and education (Patterson et al. 2004). Handover also takes place within a wider clinical context, where the communication partners are subject to different needs and priorities that impact on how they approach handover (Sujan et al. 2014). In order to make sustainable progress with improving handover, the influence of clinical systems, organisational processes and the wider institutional context have to be understood and considered.

The aim of this chapter is to illustrate the application of a systems approach to the improvement of clinical handover in emergency care. The next section describes

P. Spurgeon et al., *Building Safer Healthcare Systems*,
https://doi.org/10.1007/978-3-030-18244-1_8

why consideration of clinical handover is so important for improving patient safety in emergency care. Then, standardisation as the most common approach to improving clinical handover is described and critiqued. Subsequently, the contribution of clinical systems, organisational processes and the institutional context to the vulnerabilities of handover is analysed. A summary of the key lessons concludes the chapter.

8.2 The Trouble with Handover

A frequently used definition of clinical handover is: '*the transfer of professional responsibility and accountability for some or all aspects of care for a patient, or group of patients, to another person or professional group on a temporary or permanent basis*' (British Medical Association 2004). There are many different forms of clinical handover including shift handover between two healthcare professionals or teams of healthcare professionals, handover between two healthcare professionals from different backgrounds within an organisation (e.g. referrals from the ED to a hospital specialty) and handover that takes place between healthcare professionals from different organisations (e.g. handover from the ambulance crew to ED staff). When thinking about improving handover, it is worth bearing in mind this diversity of clinical handover because different types of handover will be affected by different performance influencing factors and specific contextual circumstances. Not all handovers are the same, and there can be significant differences.

Over the past ten years, a wealth of research evidence has been produced, which repeatedly finds that communication failures and poor handover practices are putting patients at risk of harm (Bost et al. 2010; Cohen and Hilligoss 2010; Patterson and Wears 2010; Raduma-Tomas et al. 2011; Riesenberg et al. 2009). Some of the consequences and adverse events associated with inadequate handover include hospital complications and increased length of stay following multiple handovers (Horwitz et al. 2006), treatment delays (Apker et al. 2007; Solet et al. 2005), repetition of assessments (Bomba and Prakash 2005), confusion regarding care (Ye et al. 2007; Sexton et al. 2004), inaccurate clinical assessments and diagnosis and medication errors (Petersen et al. 1994) and avoidable readmissions and increased costs (Joint Commission Centre for Transforming Healthcare 2010). In emergency care, the risks of poor handover practices are particularly important because the patients are often in critical condition, there can be several handovers within a short period of time and EDs are pressured environments with frequent situations of overcrowding (Apker et al. 2007).

We know from the literature and the investigation of adverse events a lot about the situations in which communication and handover failures occur. Early studies on handover found that frequently, there were no guidelines for how handover should be conducted, and as a result practices varied widely (Catchpole et al. 2007). Often, handover was given depending on personal preferences, with some healthcare professionals preferring a quick handover, others requesting a handover supported by notes, sometimes away from the patient and sometimes at the patient bedside. Junior clinicians in particular were expected to adapt to the preferred style of senior clinicians, and they might have to give or receive handover in different ways depending on the colleagues they were working with. Studies of the quality of documentation used during handover also suggest that documentation can be incomplete and ambiguous, and that clinicians might rely too much on the accuracy of the notes (O'Leary et al. 2010). Such performance influencing factors are exacerbated by a lack of organisational priority on handover and by the limited focus on clinician communication skills during education and training (Johnson and Arora 2009; Gobel et al. 2012).

8.3 The Benefits and Limitations of Standardisation

The most frequently suggested improvement to address the troubles with handover is standardisation (Catchpole et al. 2010; Farhan et al. 2012; Haig et al. 2006). This can be in terms of establishing specifically allocated times at which handover takes place, having a specific location for handover, and the introduction of communication protocols to structure the handover conversation. Communication protocols have been promoted widely and are known by their various acronyms, such as Situation, Background, Assessment and Recommendation (SBAR) and Age, Time, Mechanism, Injury, Signs, Treatments (ATMIST). Table 8.1 describes the handover of a patient from the paramedic to the ED consultant in the resuscitation area of the ED. Even though no formal ATMIST handover tool was used, the flow of the conversation intuitively follows the structure of ATMIST. However, even with this simple example, it is clear that handover is a dialogue, rather than a unidirectional communication.

Standardisation is an intuitively appealing intervention that has been brought into healthcare from other industries (Gawande 2010). In industries such as aviation and manufacturing, the use of standardised checklists is common and accepted practice. At first glance, standardisation of handover makes a lot of sense. It was noted above that there is significant variation in the way handover is given, and this can cause communication problems, confusion and frustration. Some time ago, I interviewed clinicians at an ED about their shift handover practices (Sujan et al. 2014), and the breadth of responses was extraordinary. In three consecutive interviews, one participant suggested that handover was much better after the introduction of a handover checklist, while the second participant declared that they had neither seen nor heard of such a checklist; and the third participant explained that they found no value in handover at all and instead came into the ED thirty minutes before the start of the shift to take a look around.

In addition to reducing such unnecessary variation, standardisation of handover can also support focussing attention on handover as a clinical task in its own right, which should be taken seriously. This can raise awareness about the importance of doing a good handover. Finally, the standardisation of handover provides opportunities for simple measurements of handover quality through clinical audits. These measurements can form the basis for further improvement.

Table 8.1 Transcript of handover from paramedic (P) to ED consultant (C)

Speaker	Transcript	Comments
P	This is [Name]. [Name] is 75 years of age. His bike has actually hit a large piece of wood in the middle of a road. He started coughing up blood, then became unconscious for approximately 3 min. Basically, on my assessment he's had a Glasgow Coma Scale [GCS] of 14. Not particularly orientated to time and place. His blood pressure [BP] was about 144/76. It is now 107	**A**ge **T**ime not given **M**echanism **I**njuries **S**igns
C	Do we know anything about the actual fall? Did he fall from the bike or did he collapse off the bike?	The consultant seeks further clarification about the **M**echanism
P	No, he fell off the bike. He hit a log in the road and he fell on his leg and his shoulder and his head. He's sustained a small hematoma to the left side of his forehead. About 3 cm laceration	The paramedic provides further information about the **M**echanism. The paramedic then describes **I**njuries
C	Did you guys give him anything?	The consultant inquires about **T**reatments
P	No, just cannulated. Airways have been OK the whole time	The paramedic clarifies that no **T**reatments have been given

However, there is a flip side to this somewhat singular focus on standardisation as the solution to the problems with handover. Even though handover has been a key patient safety priority for years, and many healthcare providers have introduced standardisation to their handover practices, there has been little actual improvement in patient outcomes (Cohen and Hilligoss 2010). While this might be in part due to the fact that our measurement systems are clearly inadequate (Vincent et al. 2008), it is probably fair to conclude that the sustainable improvement of handover remains a stubborn problem.

Considering the limited success that the main approaches to the improvement of handover have had, researchers have started looking at other ways of framing the problem with handover (Cheung et al. 2010). A fundamental criticism of previous handover research is that it is based largely on a simplified and simplistic view of how handover takes place—a kind of *work-as-imagined* (WAI) (Sujan et al. 2015b). WAI represents the views on clinical activities by those who design and manage clinical work. It expresses their beliefs about what happens in practice. Often, these beliefs are inaccurate, have gaps and more often than not represent simplifications. *Work-as-imagined* can be contrasted with *work-as-done* (WAD)—which is how everyday clinical work is actually carried out (Hollnagel 2014). Studies that attempt to capture WAD retain the richness of handover as a social activity with all its tensions, contradictions and necessary trade-offs (Sujan et al. 2015c).

While it is neither necessary nor possible to achieve full alignment between WAI and WAD, it is important to understand the gap that exists between the two views on handover. Improvements that are based only on WAI, and focused predominantly on standardisation, run the risk to mistake this view for what actually happens. As a result, improvements address only part of the problem or, in the worst case, introduce new constraints on clinical work, which can lead to workarounds and underground (i.e. undisclosed) adaptations by frontline staff (Perry and Wears 2012).

8.4 The Influence of Clinical Systems, Organisational Processes and the Institutional Context

The systems-based approach aims to understand and to describe WAD, which then forms the basis for suggestions for improvement. In Chap. 3, we introduced hierarchical task analysis (HTA) as a technique to represent clinical work. Figure 3.1 is a simplification of the handover from the ambulance crew to ED staff for the purpose of this book, but we can use it as the basis for the discussion in this section.

8.5 Work-as-Done: The Goals and Functions of Handover

Typically, a task analysis is produced in collaboration with frontline staff, for example, in the form of focus groups or mapping workshops, supported by non-participant observation. It is worth noting that the graphical representation, which makes up the HTA, is still a simplification. The analysis usually captures a wealth of rich data, which is documented in supplementary notes (e.g. scenarios and case descriptions, observation memos, etc.) that are used in conjunction with the HTA graphical representation, i.e. the analysis aims to describe *work-as-done* as fully as possible.

The analysis can provide a detailed representation and understanding of how handover is linked to clinical practice, and the different goals and functions it can serve. In Fig. 3.1, task steps 1, 3 and 4 all entail communication of patient-related information. Such communication takes place between healthcare professionals with different backgrounds and potentially different goals. For example, the handover between the paramedic and the senior ED nurse in the major area (task steps 4.2 and 4.3) involves potentially conflicting goals. The goal of the paramedic is to communicate relevant clinical and psychosocial history. The goal of the receiving ED nurse, on the other hand, is to assess the criticality of the patient and to determine the impact on the resources available in the department. Awareness of these diverse goals supports the interpretation of frequently encountered problems with handover. For example, paramedics might suggest that nurses are not listening, and nurses might feel that paramedics are rambling on. Looking at it from a systems perspective, we can con-

Table 8.2 Functions of handover

Handover serves different goals and functions, e.g.:	The people involved in handover may have different and not necessarily overlapping goals.
Management of capacity and demand	– Anticipation of demand
	– Logistics and management of demand
	– Monitoring and oversight of demand
Responsibility and delegation	– Transfer of responsibility for patient care
	– Delegation of aspects of care
Information transfer	– Communication of *immediately* relevant clinical information
	– Communication of clinical and social information
	– Archival function
Drawing attention to specific aspects	– Prioritisation of patients or information
	– Highlighting aspects of care

clude that the problem might really lie in the incompatibility of the respective goals rather than in lack of attention or inadequate communication skills.

Table 8.2 provides a summary of the different goals and functions that handover in emergency care can serve (Sujan et al. 2014). When analysing handover and the problems with handover, it can be useful to consider the goals the respective communication partners each are trying to achieve. The goals and functions are determined and set by the clinical systems (e.g. division of responsibilities for patient care between ambulance service and ED), organisational processes (e.g. handover policy and information flow within the ED) and the institutional context (e.g. national funding arrangements for EDs) within which clinical work takes place. Mismatches or misalignment in goals can give rise to tensions and problems with handover.

8.6 Systematic Identification of Major Vulnerabilities—SHERPA Analysis

In Chap. 3, we also introduced the systematic human error reduction and prediction approach (SHERPA), and a simple example was presented in Table 3.2. Just as with HTA, SHERPA is undertaken with a group of stakeholders, who can provide insights into how clinical work actually unfolds.

Table 8.3 gives examples of significant risks of handover failures in the ED that might be identified through the SHERPA analysis. The risk description gives a narrative account of the identified risk, which includes description of the failure, the potential consequences, reasons for the failure, a qualitative characterisation of the likelihood of occurrence and possible ways of mitigating the failure.

Table 8.3 Examples of significant risks of handover failures identified from SHERPA analysis

ID	Failure	Risk description
01	Full story not communicated during handover from paramedic to assessment nurse	The handover from paramedic to assessment nurse for patients in the major area of the ED is the only opportunity for paramedics to communicate verbally not only the immediately relevant clinical history, but also social information and other issues they feel require highlighting. It is also an opportunity for the nurse to ask clarifying questions and to seek additional information. If important information is not communicated successfully at this point, the nurse may underestimate the patient's acuity or may miss information such as known allergies. The reasons for this may be numerous: there may be interruptions, the nurse may not be listening with full attention, queues may lead to rushed handover, there may be too much information given by the paramedic or too little and the environment may be noisy and busy. This is a regular occurrence and might happen everyday. Possible mitigations include a dedicated handover point away from noise and queues, healthcare assistant support to the assessment nurse and a system that ensures availability of a senior clinician during handover (rapid assessment consultant)
02	Delay in handover from paramedic to assessment nurse	In those cases where the ambulance crew are queuing and waiting to hand over a patient to the assessment nurse there may be the possibility that the patient deteriorates while in the queue, potentially requiring more intensive treatment later on, for example, sepsis that is treated with delay. Causes for this are ED overcrowding or observations that have not been rechecked in the queue, possibly because ambulance crews have handed over to another crew who are now looking after several patients. Delays can happen every day
03	Patient Report Form unavailable during assessment of patient	The Patient Report Form (PRF) filled in by the ambulance crew is a detailed and comprehensive document that provides a lot of valuable, additional information to the ED staff, which is not possible to communicate during the verbal handover. If this document is unavailable, there may be a delay in assessment and treatment while the nurse or clinician look for the form, or the assessment may not be based on all available evidence, and may therefore be less accurate. Reasons for this are that the PRF travels around the ED and may get misplaced or lost as somebody picks it up but does not return it to the designated area. This can be a daily occurrence. A possible mitigation is the introduction of an electronic PRF

Why does Table 8.3 look different from Table 3.2? It is simply a way of aggregating and presenting the results of the analysis in a more reader-friendly way for the purpose of communication. The SHERPA template in Table 3.2 is predominantly for the purpose of analysis, whereas the representation in Table 8.3 is useful for communicating the findings.

The SHERPA analysis can reveal a significant number of potential risks, and this can be overwhelming. However, the risk prioritisation can help to focus attention on the most significant risks. In addition, by looking deeper into the role of clinical systems, organisational processes and the institutional context, many of the risks might actually be originating from or be influenced by similar performance influencing factors.

From the findings of the SHERPA analysis, it appears that many of the error modes of handover in emergency care are linked causally to capacity and resource issues. Inadequate patient flow may lead to overcrowding in the ED giving rise to several potential handover failures: delays in ambulance crew handover, more difficult prioritisation decisions and inadequate patient transfer handover due to unfamiliarity with the patient. In addition, in order to manage better the patient flows, handover from the ambulance crew may be taken by a senior nurse with an overview of capacity of the whole ED. The senior nurse has different information needs (see Table 8.2), which may result in information such as social history not being communicated or not being consciously heard. Inadequate patient flow into the hospital further contributes to overcrowding. This may be caused by resource constraints on the wards themselves.

8.7 System Changes to Improve Handover

While the troubles with handover are experienced and managed by healthcare professionals at the individual level, the HTA and SHERPA analyses suggest that solutions should also be targeted at the wider clinical system, the organisation and at the institutional or policy level. Competing organisational priorities, such as the management of patient flows and time-related performance targets, institutional pressures for cost savings, demographic changes, and intra- and inter-organisational models of care all impact on the quality of handover in emergency care.

There is a need for greater collaboration across departments and organisations. General practitioners, ambulance services, EDs, hospitals and other health and social services need to commit to work as partners and establish a culture of integrated and patient-centred care. It is unlikely that pressures on EDs will disappear any time soon, but efforts at nurturing relationships in order to maintain trust and respect might contribute to sustainable improvements in handover by allowing individuals from the same as well as from different organisations to understand better each other's goals and constraints.

Such system changes might appear more abstract and more difficult to achieve than the introduction of a standardised handover protocol. However, there is a growing recognition that too many of the numerous quality improvement efforts in healthcare have focused too narrowly on small fixes (such as standardisation) that fail to address problems in a sustainable way (Dixon-Woods and Martin 2016). In addition, as most of the local quality improvement initiatives are not coordinated and do not share the learning that is created, there is also the risk that at the systems level, quality is degraded rather than improved (Dixon-Woods and Pronovost 2016).

8.8 Summary

This chapter described how the application of a systems approach could generate deeper insights into the problems with handover in emergency care. Traditional quality improvement initiatives often focus on standardisation. Standardisation can reduce unnecessary variation of clinical practice and help with providing a focus on handover as a problematic issue. However, standardisation often addresses only part of the problem with handover, and frequently, the improvements seen initially are not sustained in the longer term.

Adopting a systems approach, we demonstrated that many of the problems with handover in emergency care are actually linked to capacity and resource issues, which give rise to inadequate patient flows, delays and competing priorities. Addressing these issues at the systems level requires the necessary improvement skills, senior level commitment and collaboration across departmental, organisational and institutional boundaries.

References

Apker, J., Mallak, L. A., & Gibson, S. C. (2007). Communicating in the gray zone: perceptions about emergency physician hospitalist handoffs and patient safety. *Academic Emergency Medicine: Official Journal of the Society for Academic Emergency Medicine, 14,* 884–894.

Bomba, D. T., & Prakash, R. (2005). A description of handover processes in an Australian public hospital. *Australian Health Review: A Publication of the Australian Hospital Association, 29,* 68–79.

Bost, N., Crilly, J., Wallis, M., Patterson, E., & Chaboyer, W. (2010). Clinical handover of patients arriving by ambulance to the emergency department—a literature review. *International Emergency Nursing, 18,* 210–220.

British Medical Association. (2004). *Safe handover, safe patients. Guidance on clinical handover for clinicians and managers.* London: BMA.

Catchpole, K. R., de Leval, M. R., McEwan, A., Pigott, N., Elliott, M. J., McQuillan, A., et al. (2007). Patient handover from surgery to intensive care: Using Formula 1 pit-stop and aviation models to improve safety and quality. *Paediatric Anaesthesia, 17,* 470–478.

Catchpole, K., Sellers, R., Goldman, A., McCulloch, P., & Hignett, S. (2010). Patient handovers within the hospital: Translating knowledge from motor racing to healthcare. *Quality and Safety in Health Care, 19,* 318–322.

Cheung, D. S., Kelly, J. J., Beach, C., Berkeley, R. P., Bitterman, R. A., Broida, R. I., et al. (2010). Improving handoffs in the emergency department. *Annals of Emergency Medicine, 55,* 171–180.

Cohen, M. D., & Hilligoss, P. B. (2010). The published literature on handoffs in hospitals: Deficiencies identified in an extensive review. *Quality and Safety in Health Care, 19,* 493–497.

Dixon-Woods, M., & Martin, G. P. (2016). Does quality improvement improve quality? *Future Hospital Journal, 3,* 191–194.

Dixon-Woods, M., & Pronovost, P. J. (2016). Patient safety and the problem of many hands. *BMJ Qual Saf, 25*(7), 485–488.

Farhan, M., Brown, R., Vincent, C., & Woloshynowych, M. (2012). The ABC of handover: Impact on shift handover in the emergency department. *Emergency Medicine Journal: EMJ, 29,* 947–953.

Gawande, A. A. (2010). *The checklist manifesto: How to get things right.* New York: Metropolitan Books.

Gobel, B., Zwart, D., Hesselink, G., Pijnenborg, L., Barach, P., Kalkman, C., et al. (2012). Stakeholder perspectives on handovers between hospital staff and general practitioners: an evaluation through the microsystems lens. *BMJ Quality & Safety, 21*(Suppl 1), 106–113.

Haig, K. M., Sutton, S., & Whittington, J. (2006). SBAR: A shared mental model for improving communication between clinicians. *Joint Commission Journal on Quality and Patient Safety/Joint Commission Resources, 32,* 167–175.

Hollnagel, E. (2014). *Safety-I and Safety-II.* Farnham: Ashgate.

Horwitz, L. I., Krumholz, H. M., Green, M. L., & Huot, S. J. (2006). Transfers of patient care between house staff on internal medicine wards: A national survey. *Archives of Internal Medicine, 166,* 1173–1177.

Institute of Medicine. (2007). *Hospital-based emergency care: At the breaking point.* Washington: The National Academies Press.

Johnson, J. K., & Arora, V. M. (2009). Improving clinical handovers: Creating local solutions for a global problem. *Quality and Safety in Health Care., 18,* 244–245.

Joint Commission Centre for Transforming Healthcare. (2010). *Improving transitions of care: Hand-off communications.* Oakbrook Terrace: The Joint Commission.

Nugus, P., McCarthy, S., Holdgate, A., Braithwaite, J., Schoenmakers, A., & Wagner, C. (2017). Packaging patients and handing them over: Communication context and persuasion in the emergency department. *Annals of Emergency Medicine, 69*(210–7), e2.

O'Leary, K. J., Thompson, J. A., Landler, M. P., Kulkarni, N., Haviley, C., Hahn, K., et al. (2010). Patterns of nurse–physician communication and agreement on the plan of care. *Quality and Safety in Health Care, 19,* 195–199.

Patterson, E. S., & Wears, R. L. (2010). Patient handoffs: Standardized and reliable measurement tools remain elusive. *Joint Commission Journal on Quality and Patient Safety/Joint Commission Resources, 36,* 52–61.

Patterson, E. S., Roth, E. M., Woods, D. D., Chow, R., & Gomes, J. O. (2004). Handoff strategies in settings with high consequences for failure: Lessons for health care operations. *International Journal for Quality in Health Care: Journal of the International Society for Quality in Health Care/ ISQua, 16,* 125–132.

Perry, S. J., & Wears, R. L. (2012). Underground adaptations: Case studies from health care. *Cognition, Technology & Work, 14,* 253–260.

Petersen, L. A., Brennan, T. A., O'Neil, A. C., Cook, E. F., & Lee, T. H. (1994). Does house staff discontinuity of care increase the risk for preventable adverse events? *Annals of Internal Medicine, 121,* 866–872.

Raduma-Tomas, M. A., Flin, R., Yule, S., & Williams, D. (2011). Doctors' handovers in hospitals: A literature review. *BMJ Quality & Safety, 20,* 128–133.

Riesenberg, L. A., Leitzsch, J., Massucci, J. L., Jaeger, J., Rosenfeld, J. C., Patow, C., et al. (2009). Residents' and attending physicians' handoffs: A systematic review of the literature. *Academic Medicine: Journal of the Association of American Medical Colleges., 84,* 1775–1787.

Sexton, A., Chan, C., Elliott, M., Stuart, J., Jayasuriya, R., & Crookes, P. (2004). Nursing handovers: Do we really need them? *Journal of Nursing Management, 12,* 37–42.

Solet, D. J., Norvell, J. M., Rutan, G. H., & Frankel, R. M. (2005). Lost in translation: Challenges and opportunities in physician-to-physician communication during patient handoffs. *Academic Medicine: Journal of the Association of American Medical Colleges, 80,* 1094–1099.

Sujan, M., Spurgeon, P., Inada-Kim, M., Rudd, M., Fitton, L., Horniblow, S., et al. (2014). Clinical handover within the emergency care pathway and the potential risks of clinical handover failure (ECHO): Primary research. *Health Services* and *Delivery Research, 2.*

Sujan, M. A., Chessum, P., Rudd, M., Fitton, L., Inada-Kim, M., Spurgeon, P., et al. (2015a). Emergency care handover (ECHO study) across care boundaries: The need for joint decision making and consideration of psychosocial history. *Emergency Medicine Journal., 32,* 112–118.

Sujan, M., Spurgeon, P., & Cooke, M. (2015b). Translating tensions into safe practices through dynamic trade-offs: The secret second handover. In R. Wears, E. Hollnagel, & J. Braithwaite (Eds.), *The resilience of everyday clinical work* (pp. 11–22). Farnham: Asghate.

Sujan, M., Spurgeon, P., & Cooke, M. (2015c). The role of dynamic trade-offs in creating safety—a qualitative study of handover across care boundaries in emergency care. *Reliability Engineering & System Safety, 141,* 54–62.

Vincent, C., Aylin, P., Franklin, B. D., Holmes, A., Iskander, S., Jacklin, A., et al. (2008). Is health care getting safer?. *BMJ, 337.*

Ye, K., Mc, D. T. D., Knott, J. C., Dent, A., & MacBean, C. E. (2007). Handover in the emergency department: Deficiencies and adverse effects. *Emergency Medicine Australasia: EMA, 19,* 433–441.

Chapter 9
Evaluation of the SCS Approach

This chapter highlights and comments upon some aspects of the independent evaluation of the SCS programme and approach.

An independent evaluation team from the University of Leicester was appointed by the Health Foundation. The remit of the evaluation team was to focus on evaluating outcomes of the Safer Clinical Systems *approach*[1] within the eight Trusts [referred to as 'award holder sites'], in contrast to the Warwick University Technical Support Team, whose remit was to manage the programme infrastructure and timetable and to work with the eight sites to train and support them in implementing the current version of the Safer Clinical Systems approach. The Warwick team were also responsible for 'capturing learning' about specific aspects of the approach and what works best in relation to the diagnostic and other tools used by the eight sites.

In spite of this distinction between the SCS 'approach' and 'programme', in reality we found that one impacted on the other. The complex demands of balancing the operational pace of the programme against the demands of the pre-agreed programme timetable and reporting system meant that the demands of one aspect sometimes affected the other. The operational pace had to take account of the differing starting circumstances of eight sites, e.g. organisational context and technical and process maturity of the site teams. This had to be balanced with the project management requirements of the funding organisation to manage the overall programme and maximise their return on investment.

For example, when there was a slowdown in the formal programme to allow a rethink on data gathering and the replacement of a key member of the Health Foundation team. This was welcomed by three of the site teams as an opportunity to reflect and reorganise, but the other five found that to varying degrees it reduced momentum which was difficult to regain.

Thus the programme running over two years [January 2012 to December 2013] in the NHS organisational context was affected at micro- [site team] and macrolevels [overall programme] by the contextual noise and instability of the service and by the changes within each of the ten sites, the Health Foundation and the Warwick team.

---

[1]Dixon-Woods et al. (2014).

P. Spurgeon et al., *Building Safer Healthcare Systems*,
https://doi.org/10.1007/978-3-030-18244-1_9

The first section of this evaluation is based on the continuous learning capture process undertaken by the Warwick team throughout the programme. Additional highlights from the external review team are also outlined.

9.1 New Perspectives on Safety

In developing a deep understanding of patient safety and risk, teams carrying out the Safer Clinical Systems programme used several sources of information and a number of techniques. Using a series of semi-structured interviews and group discussion, we evaluated the changes in belief, attitudes and confidence in understanding patient safety and also sought responses in answer to the question 'What approaches to proactive safety were most useful?'

Overall, the depth of knowledge and the confidence in the teams had improved—but of particular interest was the value placed on the techniques and sources used to inform safety knowledge. The usual sources used in evaluating safety included incident reports, serious incident analyses and regulators' reports. In practice, none of these was judged as being as useful for informing a proactive approach to safety as the novel techniques provided on the programme. **Process mapping, task analysis, failure mode and effects analysis** provided the most information to guide improvement.

The use of systematic tools has been common practice in other industries and is beginning to gain ground in healthcare. The new perspective is the move from a reactive to a proactive safety management system.

9.2 Applying the Learning from the SCS Approach—Some Practical Advice

We would make three broad observations on the learning from the overall project. Firstly, the five steps refer to five key sets of task or activities that need to be undertaken rigorously and in sequence, while recognising the iterative nature of the process. Secondly, the duration is entirely dependent on the particular context and circumstances within which the safety problem exists. Thirdly, contextual and circumstantial factors will also include the baseline level of relevant knowledge and expertise of the team undertaking the work.

Additionally, we learned the importance of building in a review process, which challenges the outputs of each key step. In the original project, these were referred to as 'Gates', i.e. we do not proceed through 'the gate' to the next step until we have challenged the initial outputs from the previous step. The rigorous use of the tools and the review process will almost certainly result in most teams finding that they change significantly, their initial assumptions about the safety issues in the pathway, thus underlining the iterative nature of the process.

Step 1: Define the safety problem, pathway or service and its context.
Gate 1: Check that the pathway is well-defined and manageable, and that the impact of the organisational context is understood.
Step 2: Undertake rigorous problem diagnosis leading to clarity on the nature of the safety problem and the identification of potential system changes (interventions).
Gate 2: Check that a full and adequate diagnosis has been undertaken and the limitations of the diagnostic process are understood.
Step 3: Appraise the options for possible interventions and select the most effective.
Gate 3: Check that the option appraisal addresses the most appropriate issues.
Step 4: Plan and design the interventions taking into account human and performance influencing factors.
Gate 4: Ensure that the design of interventions is appropriate and each one fully addresses human factors and performance influencing factors.
Step 5: Implementation of system improvement cycles to reduce risk and embed system changes.

9.3 The Use of the Tools and Techniques

The use of a range of tools to ensure rigorous diagnosis of the problem *before* action to resolve it and to test any existing assumptions about the nature of the problem and/or its solution is critical to an effective approach to improving safety which is based on a proactive identification of the risks in the current clinical processes. Decisions about solutions needed to be evidence based. Many NHS staff may well be familiar with some or all of these tools. The Safer Clinical Systems manual describes these in more detail.[2]

9.4 Process Mapping

"Step 2 of the Safer Clinical Systems programme is *system diagnosis*. In Step 1, you will have defined your pathway and set its boundaries, begun to understand the organisational and cultural context you are working in and identified any existing data from your current risk management systems. In Step 2, you will undertake a detailed diagnostic assessment of the pathway, beginning with a high level and then a detailed process map before going on to analyse where and how we can improve the

[2]Cooke et al. (2016).

system. After mapping, you will go on to look at failure modes, risk and your current risk control measures, the tasks people do and the factors that influence performance.

Process mapping is a powerful tool to help you understand a patient pathway or the process of care but also to achieve a common understanding and help in building a coherent team view. It is a graphical tool—it produces a visual representation as an aid to understanding, sharing information and formulating ideas for change. Based on our experience the overall steps involve:

- Get the right people in the room.
- Set out the ground rules, objectives and ethos.
- Construct a high-level process map of 6–12 steps.
- Work together to construct a detailed process map.
- Verify accuracy with other stakeholders and update accordingly.
- The construction of a comparison pathway based on participant and non-participant observer method, including following the patient journey throughout the 24 h period where appropriate.
- A comparison and gap analysis of the two pathways.

It is important to make every effort to involve all staff actually engaged in the pathway being mapped to ensure that it is accurate and represents reality, not an assumed reality. Consider whether you need to actively engage with the wider care/practice environment and staff related to the pathway.

You may face practical issues with getting all staff in the same room at the same time. Previous teams have utilised a range of 'come-and-go' or rolling focus group meeting(s) to allow the greatest number of frontline practitioners across the pathway to comment on:

- The current pathway accuracy
- Reasons why 'what is believed to happen' and 'what does happen' differs, where this has been demonstrated to reflect reality
- To validate and contribute to the fine-tuning of the 'aspired to' pathway.

9.5 Failure Mode and Effects Analysis (FMEA)

Having undertaken a detailed diagnostic assessment of the pathway, beginning with a process map, in this step—FMEA—you will go on to systematically identify how things go wrong within the existing pathway.

FMEA is a systematic analysis of process to identify the possible ways it might fail (i.e. failure modes). You can use it to examine the effects or results of failures and also the possible causes. FMEA originated in high-risk industries and has been employed in sectors such as automotive, aviation and railways for many years. More recently, it has been adopted in healthcare and is now promoted by patient safety bodies, such as the NPSA and the Veterans Affairs (VA) in the USA.

FMEA relies on the input from all people involved in the care process so, as with process mapping, application of the tool begins with getting all the right people together, clarifying the ground rules and process to be followed. In this way, it continues the development of a shared vision and a culture of safety.

FMEA is a proactive technique that is most often used to identify and address problems before they occur. In comparison, root cause analysis (RCA), another technique used in analysing errors, is used retrospectively to address problems after they occur.

Step-by-step guide overview:

- Get the right people in the room.
- Set out the ground rules, objectives and ethos.
- Review the process map.
- Work together to identify for each step how things can go wrong.
- Systematically rank the risks to patients.
- Identify possible causes.
- Record your findings.

The FMEA process can feel somewhat long-winded, and that it overlaps some of the activities required in the HTA process (see below).

There are many reasonable ways of achieving validation of the FMEA process using a range of approaches to involving staff, where getting all staff into one room at the same time is simply not feasible. Difficulties with staff engagement and/or participation were a consistent theme of the original project, not as a consequence of disinterest but as a consequence of frontline patient care demands. It may be necessary for a core team to undertake the bulk of the FMEA work and then seek to validate their perspectives as best they can with frontline staff using drop-ins, email, cajoling, come-and-go sessions.

In undertaking the risk assessment of perceived 'high-risk' points within the pathway, most teams used either the Safer Clinical Systems risk assessment tool or one already established in their organisation.

9.6 Hierarchical Task Analysis (HTA)

This system-based human factor analysis is based upon understanding the goals of the system, the tasks and subtasks needed to accomplish these, the possible failures and risks involved and the factors that cause them. The starting point of the analysis is a hierarchical task analysis or 'HTA'. HTA is a goal-driven method for documenting a process by breaking complex sequences into discrete tasks and subtasks.

These tasks can then be analysed individually for failure modes and risk as with process mapping and FMEA. An HTA has some additional advantages, however. It allows you to develop the activity up to the level of detail that is required for your purposes, with special attention to the actual task details—the things that people actually do. This in turn forms the basis of an analysis of the factors that affect the tasks—the 'performance influencing factors', or *PIFs*. By examining this in detail, we can begin to identify and manage human error. These data all feed into Steps 3 and 4 (*option appraisal and planning*) and help develop ways to improve safety and reliability of the pathway.

Step-by-step guide overview:

- Get the right people in the room.
- Agree out the goal of the step you are analysing and the boundaries of the analysis.
- Build a hierarchical task analysis (HTA) by identifying subtasks necessary to achieve the goal.
- Carry out risk ranking of each task and subtask to identify **where** things go wrong.
- Identify failure modes of **how** things go wrong.
- Analyse contextual and task factors that influence **why** things go wrong.
- Record your findings.

This tool is clearly considered to be worthwhile for the NHS in achieving clarity about tasks and processes. It also seems to be a tool that will provide transparency about precisely what is required to deliver high-risk activities within a care pathway safely. It seems that for the principles of HTA and FMEA to be optimised within the NHS, there is a need to build HTA and FMEA principles into the formulation of all clinical guidelines and pathways. This would require a complete change in the way that clinical policies are currently written and formulated.

Currently, many guidelines and pathways assume a certain level of knowledge and understanding in the professionals working with a guideline, policy, etc. What the Safer Clinical Systems project seems to have highlighted is that such assumptions are misplaced and there is not a holistic understanding of what precisely ought to be done, by whom and in what order across the professional groups engaged in a pathway. It was noted that organisational teams in the original project did not utilise pre-existing policies, procedures or guidelines in the construction of their process map, FMEA or HTA, either because no documented pathway existed, or, because what was written as 'policy' was excessively long or written in an impenetrable style. It was also notable that these teams did not seem to utilise pre-existing and quantifiable data about where things are going wrong in the pathways chosen, for example, data from adverse incident reports, clinical audits, etc. Most pathways and FMEAs were constructed based on staff beliefs and reported experiences.

The question of when to stop the HTA did seem to cause some anxiety, i.e. when to stop describing the process in greater levels of detail. It is important that teams are able to defend their decision to 'stop' the HTA process.

HTA was, at least initially, perceived as a 'complex and academic'. This meant that the training needed to be pragmatic and include activities that facilitate the building of a HTA around every day events. The lack of familiarity with the tool and lack of familiarity with using structured risk management tools (proactive and reactive) meant that participants in the original project teams did not always have sufficient knowledge and skills to merge tool elements where this was pragmatic, to avoid unnecessary repetition between the FMEA and HTA processes. But overall the HTA tool was valued by the project sites once they had got to grips with it.

9.7 Option Appraisal

At this point in the process, teams would have established the known hazards in the pathway, what is currently being done to mitigate them and how certain they are of the facts. They would also have ranked these risks in terms of severity and frequency and developed a map of the future state and the goals you want to achieve. Taking all this into account, the requirement is now to consider options for interventions. This requires consideration of what actions could be taken to minimise the present risks and how feasible these interventions are. Important points to consider included:

- The options should emerge from the work of Steps 1 and 2 and the application of the tools.
- Options should address risk as a first consideration.
- Options need to include contextual change as well as process change.
- Appraisal should consider impact (both on pathway and more widely) and feasibility.
- When considering options, the results of the Safety Culture Index (SCI), if used, should work with other information to help understand the context in which the change is being introduced and guide not only the selection but the process of implementation. The aim should be to create a context that supports safer practices not simply an improvement in the SCI indices.

Doing this in a systematic way was found to lead to a more robust plan for implementation.

Warning:

- The imperative to action, to 'do something' to address the risk, or to be seen to be taking action due to organizational expectations, must be resisted until due consideration has been given to the action needed. [Recognising that the pragmatic realities of the context may need to be managed, e.g. quick wins that will not prejudice future more fundamental action or change.]
- Going for the easy to do rather than effective, i.e. those actions which address the fundamental problem.

The following three-part process was used by the teams in the original programme.

9.8 Choice of Intervention Shortlist

- It was found to be very helpful to create a narrative of the rationale for the shortlist of proposed interventions, including how the sustainability of the intervention has been taken into account. It was important to explain how the diagnostic work undertaken in Steps 1 and 2 led to this intervention and how it was believed the intervention would take the organisation nearer to their desired future state in a sustainable manner.

9.9 Evidence to Support Decision-Making

- For each intervention, then create a narrative of the pros and cons of undertaking it.
- Consideration was given to factors including:
  - Potential impact—on pathway and on whole system
  - Feasibility
  - Ease of implementation
  - Financial consequences
  - Fit with organisational objectives and priorities
  - Chance of sustaining change.

9.10 Final Choice of Interventions

Making the final choice following discussions with key stakeholders was found to be important to get buy-in and support both from those directly involved in implementation and those who can enable wider system changes that allow a specific point intervention to take place, e.g. the Executive Lead for patient safety in the organisation, if there is one, senior managers and clinicians and the extended pathway team members that may have had less direct involvement in the development of the interventions.

The option appraisal process proved valuable. Experience showed that in addition to the six criteria outlined above, additional factors that needed to be considered were:

- Setting out a clear statement of the 'end game', e.g. 'we need to achieve a situation where…' Focusing on the position one wants to achieve can help in streamlining and prioritising solutions. Otherwise, it is possible that your vision might be limited only to the risks you want to reduce rather than the changed situation you want to achieve.
- Considering the 'reliability/consistency' attributes of the solutions under consideration. Most interventions will fall broadly within one of four types of solution:
  - Physical (designing out the human element)
  - Natural (using distance time and space)
  - Human action (over reliance on individuals and human behaviour)
  - Administrative (policies, procedures and training programmes).

Although the option appraisal process was embraced by all teams, the aspect that appeared to present the biggest challenge was 'what to leave out'. Using a structured and detached approach to this using a numerical ranking system helped. Nominal Group Techniques might also have been a method that would have enhanced the objectivity of the process.

Some teams said that more time should be allocated to test out options if the intent is to deliver more robust and consistently followed clinical pathways and thus advances in safety. To 'test out' preferred options to ensure that they are delivering their intent is eminently sensible and ought to be a core component of the action implementation processes if at all possible.

Further development of the option appraisal process could include grading 'options' according to the inherent nature of the solution suggested, e.g. 'human action', 'administrative', 'natural' and 'physical'. If the aim is to deliver enhanced sustainability and enhanced consistency/reliability, consideration of these attributes must occur. Human action interventions (i.e. instructions to staff telling them what to do) are on their own recognised as unreliable in terms of achieving day-to-day consistency/reliability in practice.

9.11 Human and Performance Influencing Factors and Related Issues

Managing the Process

Any successful process of change and improvement requires consideration of how to reconcile three potentially competing objectives:

1. Ensuring rigour and accuracy in diagnostics to ensure the right problem has been identified and understood—which requires slowing down and challenging all assumptions.
2. Maintaining interest and involvement through seeing progress and improvement—which requires maintaining momentum.
3. Ensuring sustainability of the change through working at involving staff from the beginning in order to embed the changes in process—which requires giving people the time and space to be involved in a high-pressure clinical environment with many distractions. Normal organisational churn and staff turnover can be inhibiting factors as key personnel change roles.

Some of the interventions may require organisation-wide influence, and the problem that needs tackling may reflect inherent long-standing cultural attitudes, both of which may be beyond the capability of the local team—sometimes referred to as 'wicked' or 'big hairy problems'.

Thus, despite careful diagnostics undertaken by the project teams, the Safer Clinical Systems interventions escaped few of the problems of implementation known to plague quality improvement efforts. The hazards identified during the diagnostics phase were often grounded in complex, deep-seated and long-standing problems.

Even when they had the appearance of being simple problems that would be relatively easy to fix, they were not susceptible to any straightforward remedy.[3]

There is the unavoidable time factor involved in implementing any change process successfully and Safer Clinical Systems are no different. It is not possible to resolve all issues instantly. Essentially, the Safer Clinical Systems programme is about getting a wide range of people to change their professional practice and behaviour, and this takes time.

Somewhat paradoxically, NHS organisations are not culturally inclined to reflect on their own culture and how it may be creating safety problems or inhibiting the system changes and consequent changes in behaviour identified as necessary to reduce the risks in the system.

Learning in itself can be an enabler of change, but the learning itself may seen as an imposition and therefore becomes an inhibitor of change and may increase resistance. This seems to be particularly so with 'soft' issues. It was once teams were in the midst of interventions and the problem of overcoming resistance that the critical importance of performance influencing factors became fully apparent.

9.12 Performance Influencing Factors

Some of the issues identified by teams are briefly reflected below. [The teams in the original project were looking at safety issues in handover and prescribing].

9.13 Induction and Coaching New Team Members

The issue here was the relentless churn within the clinical pathway teams. How is the understanding which supports the analysis and intervention passed on with such rapid turnover? Induction is usually focused on the 'mandatory items' and there is little time, coherence or energy for sharing the story (and learning) so far. Who is responsible and how much is enough?

This argues for stable clinical teams with clear responsibility in respect of familiarisation within induction but difficult to maintain in practice. If you were part of the initial design/diagnostic process, it is clear why you are looking for a particular set of behaviours to deliver the changes. If you joined the team later on, that awareness is missing, and there does not seem to be time set aside for this.

---

[3]Dixon-Woods et al. (2014).

9.14 How Junior Doctors Prioritise Activity

Junior doctors are major contributors to patient pathways, e.g. handover and prescribing, and they are under pressure from all sides. Without understanding how they prioritise their time and activity within a shift, the interventions designed in the classroom may not transfer well to the shop floor. Diary cards completed by junior doctors were used in some cases to build data on this. However, there was some resistance to doing this even though it would help clarify a fundamental issue.

9.15 Goals for the 'Board-Round'

A board-round is a ward round away from face-to-face contact with the patients, usually in the office using the white board or patient list. Those teams focusing on handover, in particular, suggested that the 'board-round' would be a fertile space for the distillation of information to be handed over. But there is little agreement on what is expected from a 'board-round' and what is the priority within it because different specialties, personalities and practices have different ideas. An apparently simple solution with complex implementation issues.

9.16 Prevailing Culture of Handover

There was a concern that handover was carried out as a semi-detached activity rather than a culturally engaged one, i.e. it is not seen as a core activity with potentially critical outcomes. This is a potential 'hazard' leading to high risk. Though there was general agreement about this, no teams had any easy solutions.

9.17 Ownership of the Change

There was a theme about responsibility without power. That is, those tasked with improving safety, being reliant on having the skills, motivation and stamina to influence, but not having the positional power to tell others 'to do'. Affecting change in such a heavily mediated environment was hard work and most felt they lacked access to 'a big stick'—which, interestingly, implies the assumption that the big stick is the preferred way to effect change—a cultural norm? Obviously, this may have been expressing the frustration of the team trying to effect change rather than fundamental belief. The nature of some organisational cultures and, some may say, the nature of the NHS culture as a whole, means that staff have learned to wait for the next 'big stick', and only do what it requires.

9.18 Spread and Generalisability

This was prompted by a discussion about an approach based on defining the overarching guidelines for handover but then allowing or encouraging different groups to develop their own applied solutions which meet best practice. This led on to the question whether 'any of the solutions were generalisable or were they dependent on undertaking some form of Safer Clinical Systems process first, to work effectively'?

Saying 'this is a good idea you should use it' tends to lead to poor implementation or underperformance. There is something about the journey which makes the 'solution' effective. Just how much of the journey do you have to share?

9.19 Single-Point Interventions

There was a general concern that 'events'—other initiatives, crises and changing priorities—seem to make small localised gains appear insignificant.

9.20 Hierarchy of Control

This came up regularly both as a potential inhibitor of 'challenge' even when couched as a clarifying question and also in terms of being able to look for solutions higher up the hierarchy.

9.21 Sustainability

In reviewing the problems of and approaches to achieving sustainable change to reduce risk and increase system reliability teams highlighted the following. Many of these issues relate to any process of change—but keep getting 'rediscovered'.

9.22 What Helps

- Good credibility needed, especially with doctors—of the process, the data, the personnel. Requires investing time to spend with doctors who are critical to the specific change, e.g. junior doctors, and specific senior staff who may be helpful and influential.
- Senior management support is essential as always.
- Local leadership—ideally clinical—is needed to make progress.

- Inclusion of those involved in making changes—more of this is possible if tools used effectively and collectively. Depends on situation and is time-consuming.
- Also has to be about culture, not just tools and data—understand the context.
- Evidencing the benefits according to the audience.
- Needs narrative and stories as well as data to measure.
- Needs to be aligned with strategic priorities in the organisation.
- Demonstrable financial benefit helps.
- Takes huge energy and a real belief it matters.
- Use quick fixes to give confidence but do not stop there.
- The rigour imposed by Safer Clinical Systems approach gives focus and clarity about what the problem is.

9.23 What Hinders?

- If the issue is not viewed as important enough—need to build an understanding of the problem—data, stories, cases
- Lack of job security for key roles—those leading the implementation and those being asked to change
- The team leading the change are not embedded in area affected by change—not one of us, not affected by the changes you are imposing
- Lack of clarity about aims and priorities—both at organisational level and for the specific initiative to make clinical systems safer
- A sense of being 'done to', i.e. not taking the time (slowing down) to allow staff to become involved and to build engagement.

9.24 Sustaining the Safety Improvement Approach

- Balancing time needed for good diagnostics with impatience for change and results
- Perception that process is too data heavy, e.g. Safer Clinical Systems itself could feel much more research than improvement-oriented.

9.25 Conclusion on Learning from the Safer Clinical Systems Programme

While some of the points raised here are specific to Safer Clinical Systems, they are rather reflective of large-scale change attempts. The uncomfortable reality is that the NHS needs to accept and overcome the resistance encountered, whether from inertia or vested interests, as the ultimate reward is greater safety for the patient (all of us at some time). It also needs to be recognised that proactive search for hazards and a focus upon risk assessment (and management) is not a familiar approach in any health system. It is exactly what is needed and hence why points of difficulty in implementation need to be recognised and countered and not viewed as reasons to avoid doing the right thing.

9.26 Some Key Points from the External Evaluation

The overall conclusion of the evaluation team was that:

As an example of a proactive, problem-sensing approach to diagnosing and treating defects in systems likely to create the conditions for harm, the Safer Clinical Systems approach should be regarded as very promising. Many of its principles are sound; some of its features can be updated in the light of learning from this programme. The approach would benefit from further use-in-practice and evaluation.[4]

In any future application of the approach, we would advocate the 5-Step approach outlined in Chap. 5, but with three qualifications. Firstly, the five steps should be seen five key sets of task or activities that need to be undertaken rigorously and in sequence [with appropriate iteration] and do not necessarily require a formal programme of the nature of Phase 2 Safer Clinical Systems. Secondly, the duration would first of all be much shorter than the Phase 2 programme, but as to the 'right' duration, this is entirely dependent on the particular context and circumstances within which the safety problem exists. Thirdly, contextual and circumstantial factors will also include the baseline level of relevant knowledge and expertise of the team undertaking the work.

The teams started with the diagnostic phase, including the pathway definition. This was seen as a distinctive strength of the programme by the participating site teams. With a few exceptions, they valued it highly. The diagnostic phase was praised by participants for its ability to make visible aspects of systems that had previously been obscure, to identify weaknesses that had not been previously recognised or understood and to challenge pre-existing assumptions.[5]

In reality, there was often pressure from Board level for the diagnostic phase to be completed quickly, particularly where risks were being identified, in order that action could be taken to 'put it right'. However, the requirement on teams to document and demonstrate evidence of identified risks before proceeding to action enabled teams to push back and give themselves space to properly understand the problem. This was enabled by being part of a bigger project with independent experts on hand to advise and support. At the same time, there was sometimes pressure from within the teams to push on faster once they had understood the problem and a degree of frustration was expressed at being forced 'to proceed at the pace of the slowest ship in the convoy'.

[4] Dixon-Woods et al. (2014).

[5] Dixon-Woods et al. (2014).

One area where several teams departed from the approach prescribed by the support team was in fixing some problems as soon as they found them, rather than waiting to carry out Steps 3 and 4. If problems identified during the course of the diagnostics were alarming in their capacity to cause patient harm, some teams felt it would be immoral and wrong to delay fixing them (especially if the solution was very straightforward). Teams also sometimes wanted to generate some 'quick wins' and demonstrate to colleagues that the project team was capable of practical action.[6]

Reality required a pragamatic response to such circumstances. The overriding consideration was whether or not 'the problem' had been rigorously and objectively identified. In terms of creating and sustaining buy-in to the work of the local SCS teams among the wider cadre of clinical staff and at Board level, quick wins were an effective aspect of the change strategy.

9.27 The Diagnostics—Some Examples of Underlying Safety Problems

The following illustrates some of the findings from the diagnostics highlighted by the evaluation.[7]

Organisational contextual pressures creating risks in clinical systems:

Many clinical systems were highly unreliable and laden with potential to harm patients. Clinical staff at the sites were typically under severe production pressures with very high workload and multiple competing priorities. Staff shortages often meant that it was not possible to ensure that systems functioned as they were supposed to.

Organisational cultures not supportive of patient safety:

Organisational and professional cultures were not always fully aligned with the goal of achieving patient safety. At some sites, for example, staff perceived that there was a blaming culture that junior doctors felt alienated and lacking in support, that roles were poorly defined, that staff tended to be highly task-focused because of the pressure of workload and that multi-disciplinary working was weak.

Not matching staff skills and experience to patient safety needs:

Reliable functioning of many clinical microsystems was also challenged because it depended on staff in training (junior doctors), non-permanent staff (locum nurses, doctors and other staff) and others whose competence, skills and confidence were highly variable.

Individualised and subjective approaches to clinical practice:

Much of the variability and associated unreliability arose because of the absence of clearly agreed standards. Consultants did not always provide the necessary leadership in taking charge of these problems, and did not effectively standardise their

[6]Dixon-Woods et al. (2014).

[7]Dixon-Woods et al. (2014).

practices—meaning that junior staff and nurses had to spend effort learning and anticipating what each one would expect.

Ad hoc accumulation of system practices:

In many cases, systems for achieving particular tasks or functions had never been purposefully designed or made explicit; instead, their practice had become accepted through repeated use.

The disabling effect of poor technical and human systems:

Communication and coordination were among the most important sources of hazards in the eight sites, poorly functioning IT systems, challenges in coordinating different professional groups to meet in one place at the same time.

The adverse impact of accumulated system inefficiencies:

The consequence of these multiple defects was that staff were often hassled and distracted by the 'small stuff'—components of systems that did not work properly and took large amounts of time to repair or rescue—and found it hard to keep the bigger picture in mind. Systems were therefore often stressful to use, created distractions or interruptions and wasted resources and time. This level of unreliability was likely to contribute to problems in assuring safety.

The use of the diagnostic tools and the requirement to use them collectively and on a multidisciplinary basis enabled all of the teams to develop a shared insight into the underlying hazards and risks within their clinical systems that had not previously existed. It enforced a discipline and rigour that was not usual in identifying patient safety problems.

9.28 The Interventions

The challenges of options appraisal meant that not all of the interventions were well-suited to, targeted on, or fully aligned with the problems that were identified during the diagnostic stage.[8]

This was an area where the demands of the SCS programme affected the SCS approach. For various reasons, there was a hiatus in the programme timetable for key learning and review events between completion of the diagnostics and the identification and selection of interventions and associated data gathering. The impact of this was further compounded by the fact that it took place at the end of the Trust financial year requiring a number of key managers to address internal budgetary issues, together with a change of key personnel at the Health Foundation.

It was initially assumed by the Warwick team that the process of identifying and then selecting appropriate interventions using the advice and guidance provided would be relatively straightforward but some site teams struggled with this. Not all interventions were clearly rooted in the outcomes of the diagnostic process and emerged from local consultations and discussions. In retrospect, the sites needed

[8]Dixon-Woods et al. (2014).

more time to both develop the interventions and then assess or appraise them in relation to their impact on the risks identified and their 'doability'.

Some interventions selected by the sites which are characterised as 'administrative' in relation to the hierarchy of control literature can contribute to quite significant improvements, even though this is regarded as a weak type of intervention.[9] In the context of the programme, the sites were under pressure to select an implementable intervention which could be subject to measurement, and had they focused on the issues which may have prompted a different categorisation in the level of hierarchy, the programme would have run out of time. This is a reality constraint which should not be overlooked.

In any future application of the approach it would be necessary to take Step 3 [option appraisal] and option 4 [planning] in a more consciously deliberate manner before proceeding to Step 5 [system improvement].

Nevertheless as is outlined in Chap. 5 describing the SCS approach, most sites achieved demonstrable shifts in reducing risks identified. The use and application of measures to demonstrate the effect of interventions was an area that the evaluation team questioned and initially it was assumed that common measures could be applied to all sites. However, given the design of the SCS approach leading to the individual nature of the problems identified in each site meant that this was never likely to be the case. Each site had to design and 'safety set' of measures that would demonstrate the changes in their particular circumstances. But data collection and management did present a number of technical and capability challenges to the sites which probably reflects a wider NHS issue.

'Doability' mentioned above is dependant on a mix of issues to do with both technical skills associated with the intervention and, often more fundamentally, the knowledge, experience and authority to effect the requisite changes in organisational systems and behaviour. The evaluation team highlighted this issue:

Responsibility for responding to the findings of the diagnostics (that is, the responsibility for supporting the improvement to happen) should lie with senior management, not with small clinical teams. Senior management should identify the level at which changes need to be made and who should make them, with what resource and authority. This will mean strengthening the problem-solving capacity of healthcare organisations from the top to the bottom.[10]

This is particularly true for problems that are rooted in the organisational and cultural context which are beyond the capacity of a local clinical team to influence. Often an apparently simple problem such as getting the right group of clinical professionals into the same room at the same time for a handover meeting is stymied by different shift patterns, hours of work, other scheduled duties, etc. that require wider system change authorised at a higher level.

[9]Liberati et al. (2018).

[10]Dixon-Woods et al. (2014).

9.29 Post-programme Response to the Evaluation and Follow-up Work

Following the conclusion of the programme and the publication of the evaluation report, the Warwick team have undertaken a number of follow-up initiatives to test and extend the application of the SCS approach.

We have developed a 'DIY' manual to assist any organisation that wanted to use the approach.[11] However, this has not been tested on a stand-alone basis, i.e. without expert support. As the evaluation team commented:

It is unlikely that the approach can be learned from a manual; the skills are likely to require specific training based on a well-specified curriculum.[12]

We nevertheless believe it important for the approach to be made accessible to everyone. [The manual can be downloaded.]

With respect to their comment on training based on a curriculum we have taken two initiatives to develop this. The first is through the medium of a Warwick University Masters level Module working with one Trust to see how easy it was to get across the approach and for busy clinical managers to apply it to a current safety issue, drug administration. This was supported by funding from the Health Foundation. The module was delivered on site to staff [multidisciplinary] in order to assess how far it was possible to convey the Safer Clinical Systems approach to safety in a short structured postgraduate module of three taught days with interspersed periods for local team working on revision and application. The programme was also evaluated [unpublished] which found a consistently positive reaction to the ideas and tools of the Safer Clinical Systems approach. Some caveats were expressed in relation to differing expectations of timescales and progress between the managers on the programme and the Board which replicates a finding from the original programme, i.e. the imperative to action irrespective of whether the action is aimed at the right problem in the right way. All participants stated they had no difficulty in getting to grips with the ideas and principles of SCS—identifying what could go wrong and taking a whole systems approach. Some comments indicated a more practical/pragmatic perspective, possibly related to the individual's role in the organisation, e.g. ward management focus versus wider pathway/service focus. There were a range of comments to the effect that using the tools such as process mapping together with the overall philosophy of the approach forces insights into the reality of what is actually taking place. Time—time to think and reflect individually, time to work together collectively was identified as the main inhibitor, again reflecting issues raised in relation to the original programme. Some individuals seemed to have more proactive support from their senior managers to do the practical work associated with the application of the learning but day-to-day clinical pressures and emergencies intervene for all.

The second is a more ambitious project. Members of the Warwick team are working with Health Education England and the Academy of Medical Royal Colleges

[11] Cooke et al. (2016).

[12] Dixon-Woods et al. (2014).

to design and develop a Postgraduate Medical Patient Safety Curriculum using the proactive approach to patient safety (Safer Clinical Systems).

Following a review of selected college curricula and follow up interviews with Royal College of Surgeons, Royal College of Radiologists and Foundation Programme, it became clear that there is no recognisable point during training where patient safety is taught as such. Patient safety is implicitly at the heart of all college curricula, but there is a clear risk that if something is implicit in everything, that it appears explicitly nowhere. When patient safety is covered, the emphasis is almost exclusively on the individual responsibility of the doctor for the patient in front of them, with very little acknowledgement of the impact of the system on safety.

Discussions with a focus group of NHS England Medical Director's clinical fellows produced very similar results, with comments about patient safety training being very diffused throughout the curriculum, with nothing on the systems aspects, but everything being about personal culpability.

9.30 In Conclusion

The evaluation team concluded that

Many of the principles underlying the Safer Clinical Systems approach are generally sound and rich in promise and would benefit from further use-in-practice and evaluation.[13]

However, we would recognise that the approach would benefit from continuing testing and adaption in practice, particularly in the area of developing and implementing targeted interventions. The current strength of the approach is in the structured use of diagnostic tools to build an objective and shared understanding of how things are actually working in practice, as against subjective, varied assumptions.

The approach would be improved by better fit between the findings of the diagnostics and the nature of the organisational response and by clearly locating the responsibility and imperative to act with senior management.[14]

The role of frontline clinical teams in effecting improvements will always be critical but should be clearly and explicitly linked to the nature of the problems and the capability and authority of the team to solve the problem at their level.

There needs to be tight and logical coupling between hazards, interventions and measures and reviewing and revising these over time.[15]

---

[13]Dixon-Woods et al. (2014).

[14]Dixon-Woods et al. (2014).

[15]Dixon-Woods et al. (2014).

References

Cooke, M., Cross, S., Flanagan, H., Jarvis, R., & Spurgeon, P. (2016). *Safer clinical systems. A new, proactive approach to building safe healthcare systems*. A reference guide for clinicians and managers. Coventry: Warwick Medical School (Safer Clinical Systems team), University of Warwick. http://patientsafety.health.org.uk/sites/default/files/resources/hf_safer_clinical_systems_reference_guide_final_1.pdf.

Dixon-Woods, M., Martin, G., Tarrant, C., Bion, J., Goeschel, C., Pronovost, P., et al. (2014, December). *Safer clinical systems: Evaluation findings*. Learning from the independent evaluation of the second phase of the Safer Clinical Systems Programme. London: The Health Foundation.

Liberati, E. G., Peerally, M. F., & Dixon-Woods, M. (2018). Learning from high risk industries may not be straightforward: A qualitative study of the hierarchy of risk controls approach in healthcare. *International Journal for Quality in Health Care, 30*(1), 39–43.

Chapter 10
Moving Forward: A New Patient Safety Curriculum

We have argued throughout this text for change in approach to patient safety—simply and bluntly because the existing models are not working. There is a building momentum to this position. Kellogg et al. (2017) assert that the current adherence to root cause analysis (RCA) as the investigative model is leading to at least two alarming consequences:

- Approximately 35% of the outcomes of RCA investigations are staff training and policy reinforcement (the weakest and least sustainable interventions).
- And that 'multiple event types were repeated in the study period' (p. 381).

There are quite well-documented deficiencies in the conduct of the RCA process (Spurgeon et al. 2017, op.cit.) identifying issues such as a tendency for a persistent blame culture to inhibit reporting to be a cumbersome and bureaucratic process and rarely addressing the major problem of the future risk.

The retrospective approach that is exemplified by the RCA process (Spurgeon et al. op.cit) is not without value, but the continued level of exposure of patients to adverse incidents (see Chap. 1) demands an advance from the status quo.

In an interesting and sophisticated analysis, Baddeley (2017) raises the awkward question of just how individual (doctor) decision-making functions within a devolved and democratised delivery system which is increasingly beyond the control and influence of the original decision maker. He is indirectly also giving voice to the investigative dilemma of 'who was to blame?' versus how did the system deficiencies facilitate the incident.

The safety approach we have advocated would indeed shift the balance of any enquiry into the systemic issues that may have contributed to an individual making an error. We have advocated a model underpinned by the discipline of human factors but going beyond the communication focus often associated with human factors. It is human factors in its wider context incorporating the proactive search for hazard and risk within any delivery system and thereby interventions to eliminate or control this risk (see Chap. 3 and the recently launched report of the Clinical Human Factors Group).

P. Spurgeon et al., *Building Safer Healthcare Systems*,
https://doi.org/10.1007/978-3-030-18244-1_10

We also advocate learning lessons from other industries who have made progress in keeping their operation safe. The aviation industry is frequently (too frequently) mentioned, and Oliver (2018) rightly points to the somewhat inappropriate nature of this comparison. However, it is the principles of the approach that we are advocating—essentially proactively seek out the sources of risk and eliminate or control prior to (take off) or a patient is harmed (see Chap. 2 for a conceptual framework of how we might integrate learning from other sectors).

Perhaps the nearest parallel to the changes we are proposing is to be found in the Resilient Healthcare Movement (RHC) described by Hollnagel et al. (2013). The key component of enhancing Safety I with Safety II has been discussed in Chap. 1. In accord with our own thinking, they argue that a systemic approach is necessary to improve patient safety. Their arguments about recognising the dynamic interactive implications of introducing a change to the delivery system are not dissimilar to our proposed use of the safety case (Chap. 6).

Safety II recognises the merits of safer design and a focus upon risk but goes further in an attempt to incorporate the dynamic complex nature of modern healthcare. It suggests that flexibility, adjustment and local adaptation by the practitioner are always required to cope with the changing circumstances on the ground. This is both appealing (offering the clinician the final say) but also a potential undermining of the approach as it introduces a variability based on the judgement of the individual clinician. As argued before in Chap. 1, this is workable with excellent job incumbents, but in all worlds of work there are variations in standards and competence. Therefore, Safety II has the potential to endorse local variation by mediocre or perhaps less able individuals who (and their patients) might be better served working within a set of system parameters rather than individual judgement.

Hollnagel et al. (2013) do though helpfully emphasise the difference between work-as-done (WAD) on the frontline from work-as-imagined (WAI) often by managerial parts of the organisation. This is again very similar to our argument that in order to have a new approach to patient safety it is essential to have a new mindset endorsed by all levels and types of staff. This we believe can only be achieved by introducing into the training of all staff both the philosophy underpinning the proactive, risk-oriented approach but also the practical tools to carry it out. It would appear that we are not alone in identifying the need for a new approach. The longevity of the sustained level of adverse events is in itself a clear, logical argument for an impetus to change. However, there would seem to be an acceptance of a cultural zeitgeist that is creating the necessary conditions for change with a number of recently published papers all advocating movement in a similar overall direction. NHS Education for Scotland (2018) has published a useful model emphasising the need for systems thinking in healthcare. Similarly, the Chartered Institute of Ergonomics and Human Factors (2018) has produced an excellent white paper advocating the value and impact the enhanced application of human factors would have in healthcare. Valuable as these publications are they represent disciplines or subdisciplines and do not yet create an overall behavioural framework that constitutes a collective, holistic approach to patient safety for all healthcare staff. Experts in human factors system thinking could have a significant impact in health settings but operating as experts they remain an influential but small cadre of staff. Our position, as represented in this new syllabus, is to provide training at appropriate levels for all staff such that the essential shift in thinking about patient safety comes to permeate the whole service.

The Institute for Healthcare Improvements (Frankel et al. 2017) has offered an important initiative in terms of such a framework. The Framework for Safe, Reliable and Effective Care covers most of the key areas with a particular emphasis upon cultural and learning variables. The concepts in the framework are expressed at quite a high level with limited behavioural advice. The enactment is allocated to the traditional PDSA cycle which is well established but as something of a universal tends to be undertaken in practice with varying levels of proficiency and comprehensiveness. Again we would suggest the framework presented here provides clear behavioural components that can be acquired continually as appropriate to the existing and future career stage.

Patient Safety Learning (an independent association with patient-based commitments to improving safety) has published a very important and comprehensive new paper. 'A Patient-Safe Future' (2018), which in itself, presents a powerful argument for exactly the sort of safety framework we have developed here. Not only do they argue for a new approach to patient safety (as we have in this text), they also suggest that 'we lack a national framework describing the skills and competencies that are needed to enable everyone – carers, clinicians, auxiliary workers, managers and board members to support and contribute the crucial patient safety related activities' (p. 8). They also argue that such a framework be capable of providing relevant training at the appropriate level for different staff groups and the tasks they perform. Finally just as we do they argue for a new senior individual responsible for an overall focus on safety in the organisation. This would have the merit of professionalising patient safety as a discipline in healthcare (as it is in other safety clinical industries).

We turn now to the development and content of this new patient safety framework and syllabus as a basis for whole service training and development. The outline presented here is the consequence of the current project. We would expect modification to occur as the syllabus is incorporated into various educational contexts. It is though we believe the first holistic, multidisciplinary approach to training for patient safety for all staff. The remainder of this chapter presents the key principles of the syllabus and the detail of the underlying competencies that can be organised into appropriate educational curriculum for the various staff groups. The material will then form the training building blocks for the new National Patient Strategy.

Patient Safety Syllabus
A syllabus for education in patient safety
A proactive systems-based approach to safety
Preventing harm before it occurs.

Printed with kind permission of the Academy of Medical Royal Colleges

10.1 Patient Safety Syllabus

About this Syllabus—What You Need to Know

Who Is It For?

- The syllabus is designed with multi-professional capability.

Why Is It Different?

- The first syllabus in patient safety includes methods for the proactive management of risk to patients.
- Moves safety from reactive systems, where we are continually looking backwards at harm, to proactive systems to prevent harm occurring in the first place.

10.2 How Will It Make a Difference to Clinicians?

- It provides content and navigation to support all patient safety activities—including incident investigation, creating a safety culture, using human factors, proactive risk management and managing human error.
- It includes capabilities that clinicians can develop and apply at all levels of their work.

10.3 Is It Just About Non-technical Skills?

- No. It includes human factors such as communication and situational awareness but has a central focus on how to create safe clinical systems.
- The syllabus includes for the first time a comprehensive human factors approach.
- This requires an understanding of the systems clinicians work within, human behaviour and human error.
- The four key themes used throughout the syllabus are human factors, systems expertise, risk expertise and safe culture.

10.4 Where Does This Work Come From?

- The work builds on previous work in the NHS on patient safety, academic courses in patient safety, the national programme Safer Clinical Systems and direct experience in managing safety in NHS Trusts.
- The development of the syllabus has been guided by an expert advisory group including representatives from academia, the General Medical Council, royal colleges, NHS Improvement and NHS England. The work was funded by Health Education England, and other jurisdictions have been consulted.

10.5 What Impact Will This Work Have?

- We believe that this syllabus will be relevant to patient safety education at all levels and in all professions.
- The syllabus is expected to create a step change in thinking about safety, moving the emphasis from reactive to proactive methods.
- The syllabus takes the systems approach to safety that has been continually advocated across the world and sets it out clearly for professional education.
- Throughout the syllabus, the emphasis is pragmatic and focuses clearly on how to build safe clinical systems in clinical areas, pathways and specialities.

10.6 Patient Safety Syllabus

Introduction

Patient safety continues to be a significant issue in healthcare and a focus of both quality improvement and academic research.

It is structured to provide both a technical understanding of safety in complex systems and a suite of tools and approaches that will:

- build safety for patients
- reduce the risks created by systems and practices
- develop a genuine culture of patient safety.

Though there are a number of well-known safety procedures in healthcare—including the intention to learn from incidents and some key national safety regulations—this curriculum is distinct in two ways. Firstly, it draws explicitly from widely used safety methodologies applied routinely in other safety-critical industries such as aviation and process engineering. These are industries where the use of a systems-based approach and the recognition of human error management have

Fig. 10.1 Key domains in safe clinical systems curriculum

brought safety to high-risk areas and have long been upheld as learning opportunities for healthcare. Secondly, and in line with best practices from safer sectors, we adopt an approach that brings a systems perspective to *reactive* safety methods and—perhaps most importantly—uses a systems approach to enhance patient safety *proactively*.

The curriculum consists of five sequential domains, drawn from key developing themes in patient safety, which we outline in the next section. Further sections in this document describe the outcomes expected and the key capabilities which will be developed. The knowledge, skills and behaviours considered effective in patient safety, and some suggested approaches to assessment are provided as illustrations for a single domain, *learning from Incidents*, in Appendix A.

Key Domains and Underpinning Knowledge

The domains of this curriculum are presented below as a linear sequence, though there are inevitable dependencies and synergies between them. To understand this, and to support the structure and content of each domain, we have set out the key outcomes for each domain and the underpinning knowledge and expertise required at each stage (Fig. 10.1).

The rationale used in developing the domains embodies a spiral of learning, with each domain building on and deepening the work carried out in previous domains. The elements of underpinning knowledge and expertise fall into four key themes that run through each of the domains and, through the unfolding of further knowledge within each domain, build a comprehensive understanding in each area. The key areas of underpinning knowledge and expertise are:

- Human factors
- Systems expertise
- Risk expertise
- Safety culture.

Though elements of each area will be used in each domain, some domains have a strong focus on two or three areas. For example, Domain 2 (learning from incidents) draws most deeply on expertise in risk expertise and human factors; Domain 4 (Creating Safe Systems) draws more from systems expertise and safety culture.

The overall structure of the curriculum also focuses on **knowledge**, **action** and **consolidation**. Thus, Domain 1 provides the systems *knowledge* which is critical to carrying out the necessary *actions* in reactive approaches in Domain 2; Similarly, Domain 3 provides the *knowledge base* for *actions* in proactive approaches to patient safety in Domain 4. Domain 5 draws on all previous domains to provide the knowledge and tools that consolidate and maintain patient safety.

Appendix B provides further details on the underpinning knowledge and expertise within each domain.

The following sections take the domains above and specify the capabilities in more detail. Each domain contains four subsections describing key capabilities, and within each subsection are more detailed capabilities to be attained in building expertise in the area. In addition to the detailed capabilities, we have provided examples of generic learning and development activities, themselves divided into those to be delivered in the early part of training and those to be mastered at a higher level.

Key to Structure

Each capability is presented with essential learning outcomes in the left-hand box, together with, in the right-hand column, examples of overall learning activities at early and higher specialist *levels*.

Capabilities	Examples of generic learning and development activities	
	Throughout postgraduate study	
	Early specialist training	Higher specialist training

Outcomes

Overarching Outcome

The overarching outcome that this syllabus seeks to achieve is that all staff, both clinical and non-clinical will be able to:

Demonstrate a clear understanding of both reactive and proactive methods of approaching patient safety, including personal clinical safety and the wider factors that impact system safety.

10.7 Domain 1—Systems Approach to Patient Safety Outcomes-(See Fig 10.2)

Demonstrates knowledge of how culture and working systems lead to risks to patients.

> Understands how systems failures create risks to patients; recognises how organisational culture can lead to failure or improvement in clinical practice; and understands and acts on national regulation and findings of national case studies in patient safety.

1.1 The Safety Landscape

<table>
<tr><td rowspan="2">1. Has knowledge of national learning reports and can describe key findings
2. Apply lessons from key case studies in patient safety in the specialty and the organisation
3. Analyse patient harm levels to evaluate the safety of the specialty and the organisation
4. Relate patient harm and evidence of near misses in patient safety to local systems design and practice</td><td colspan="2">Applies key learning in patient safety to the local environment</td></tr>
<tr><td>Supports system evaluation and improvement in the organisation</td><td>Leads system-based evaluation and improvement in the organisation</td></tr>
</table>

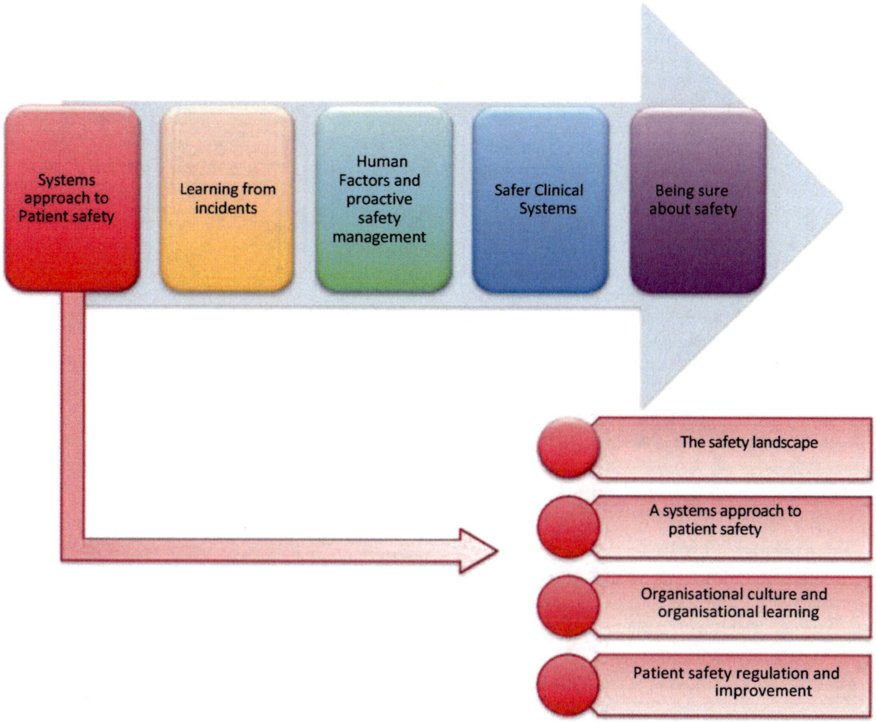

Fig. 10.2 Four key elements in Domain 1

1.2 System Approach to Safety

1. Recognise and describe the effect of systems design on risk and safety 2. Outline the principles of direct and latent failures and of performance influencing factors 3. Describe safety approaches used in other safety-critical industries 4. Explain the fundamentals of human factors and human error	*Actively applies an understanding of systems to improving safety in the specialty*	
	Develops the ability to understand safety as beyond safe individual practice	*Ensures that system risk and clinical risk are both addressed in improving safety in the specialty*

1.3 **Organisational Culture and Organisational Learning**

<table>
<tr><td rowspan="2">1. Recognise organizational culture and the principles of safety culture
2. Explain the effect of blame culture on organizational learning
3. Lead group work on safety culture in the specialty
4. Analyse and evaluate safety culture and organizational learning within the specialty</td><td colspan="2">Uses an understanding of organisational culture to identify and improve patient safety</td></tr>
<tr><td>Support individual and group work to evaluate safety culture</td><td>Leads on developing a safety culture within the specialty</td></tr>
</table>

1.4 **Patient Safety Regulations and Improvement**

<table>
<tr><td rowspan="2">1. Outline and explain key safety recommendations from professional bodies and regulators, including mandated safety practices
2. Evaluate and ensure that recommendations are complied with within the specialty
3. Identify key areas where safety can be improved within the specialty
4. Act to ensure that improvement initiatives are in place and monitored within the specialty</td><td colspan="2">Ensures that key safety and compliance data are monitored and subject to improvement</td></tr>
<tr><td>Monitors safety data and identifies improvement areas or non-compliance</td><td>Leads on creating full compliance with safety measures in the specialty</td></tr>
</table>

10.8 Domain 2—Learning from Incidents-(See Fig. 10.3)

Outcomes

Conducts a System-Based Investigation into Patient Safety Incidents, Treating Individuals Fairly and Creating Future Safety

Uses a system-based approach to investigating patient safety incidents; understands and addresses human error in incident investigations and responses; and distinguishes between system-based failures and failures in individual performance.

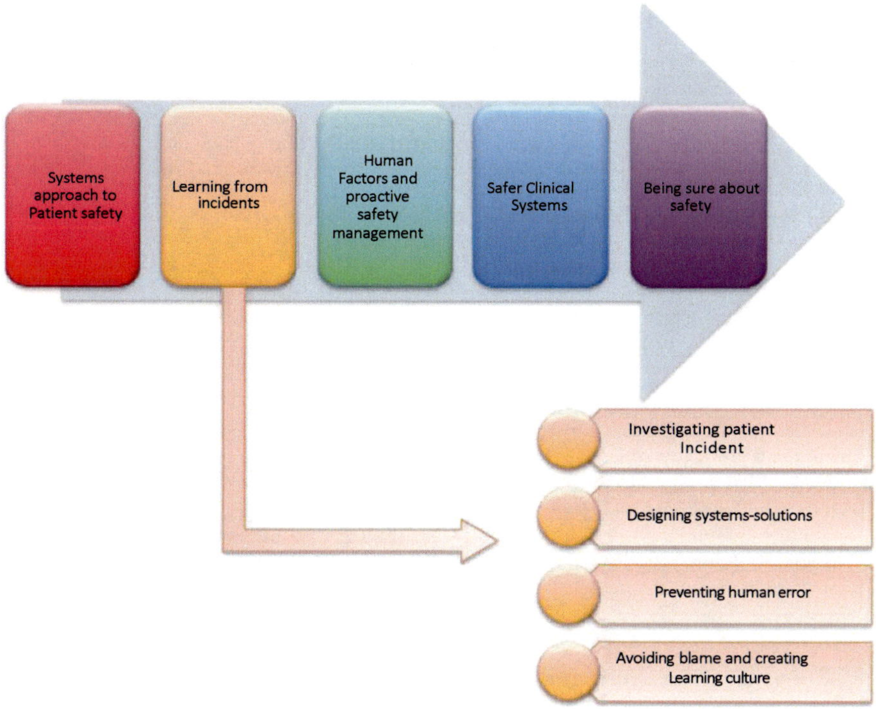

Fig. 10.3 Four key elements in Domain 2

2.1 Investigating Patient Safety Incidents

<table>
<tr><td rowspan="2">1. Ensure that a multidisciplinary team steers the process
2. Create an evidenced timeline for the patient journey through document review and unbiased interviewing
3. Use a systematic approach to identifying causal and contributory factors in analysing incidents
4. Use an understanding of human error to describe discrete care delivery problems
5. Ensure that incident reports include clear recommendations for change</td><td colspan="2">Responds to patient safety incidents to improve future safety</td></tr>
<tr><td>Takes part in the system-based incident and near miss investigations</td><td>Leads system-based incident and near miss investigations</td></tr>
</table>

2.2 Designing System-Based Solutions

<table>
<tr><td rowspan="2">1. Use the wider system and context to respond to incident investigations
2. Use an understanding of each separate care delivery problem to bring about changes in the system which will prevent future harm
3. Use awareness of stronger and weaker interventions when developing safety interventions
4. Check the robustness of interventions for the impact on future risk and safety</td><td colspan="2">Uses an awareness of systems factors to reduce risk to patients and improve safety</td></tr>
<tr><td>Contributes system-based thinking to incident investigations</td><td>Carries out investigations that lead to safety into future clinical systems</td></tr>
</table>

2.3 Preventing Human Error

<table>
<tr><td rowspan="2">1. Ensure that incident investigations recognise and highlight human contributions to risk and patient safety incidents
2. Apply an understanding of error as a consequence of systems rather than an explanation of safety failures
3. Evaluate human error to design effective safety interventions
4. Build human error management explicitly into incident investigation reports</td><td colspan="2">Uses an understanding of each separate care delivery problem to bring about changes in the system which will prevent future harm</td></tr>
<tr><td>Uses an understanding of each separate care delivery problem to bring about changes in the system which will prevent future harm</td><td>Uses an understanding of each separate care delivery problem to bring about changes in the system which will prevent future harm</td></tr>
</table>

2.4 **Avoiding Blame and Creating a Learning Culture**

1. Explain how to distinguish between system-based failure in safety and the contribution of individual clinicians 2. Use the 'Just Culture Tool' (JCT) with each individual failure in a systematic way to challenge and validate individual behaviours 3. Document and share the outputs from the JCT with those involved in the incident and the investigation to ensure complete transparency 4. Demonstrate that systems failures identified by the JCT and addressed in the response to the incident or near miss	*Accepts the existence of individual errors and contributes to a culture of sharing presentative interventions*	
	Contributes to understanding of individual error in investigations	*Leads in sharing briefings from incidents and near misses widely in the organisation*

10.9 Domain 3—Proactive Management of Patient Safety-(See Fig. 10.4)

Outcomes

Evaluates and Ranks Risks to Patients in the Systems and Culture of the Specialty

Understands and is able to categorise tasks and their risks in clinical practice; recognises the impact of non-technical skills; and uses measures of process reliability to monitor and improve safety.

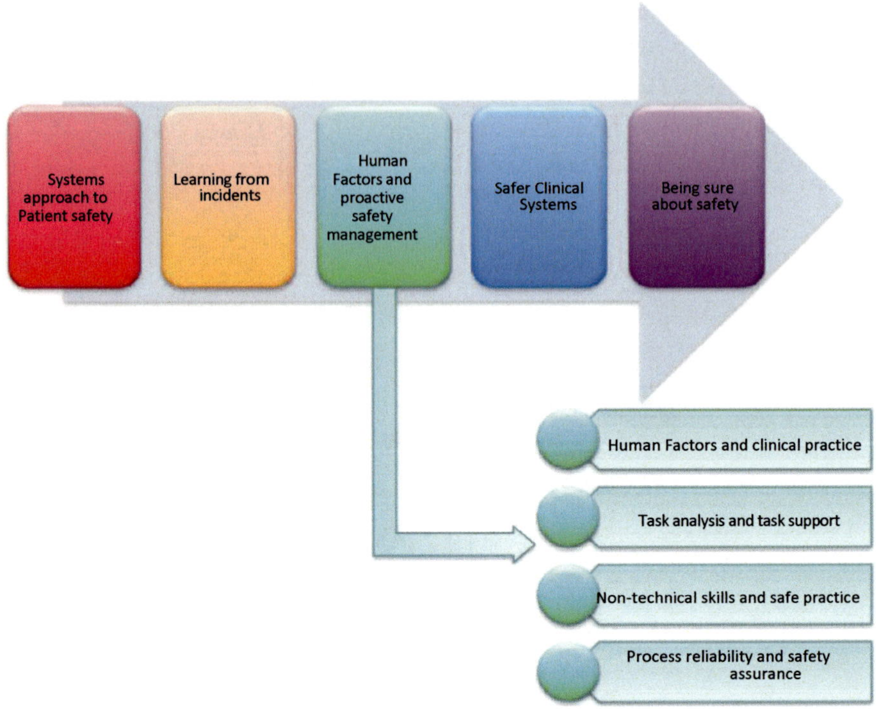

Fig. 10.4 Four key elements in Domain 3

3.1 **Human Factors**

<table>
<tr><td rowspan="2">1. Outline and explain the role and effect of humans in complex systems and the fundamentals of human factors
2. Reflect on specialty performance to explain human factors in practice
3. Evaluate the key factors that affect human performance and relate
4. Demonstrate knowledge of the effect of human factors management in safety-critical industries</td><td colspan="2">Develops an understanding of human performance in clinical systems</td></tr>
<tr><td>Recognises and accepts the limits of human performance and the effect on clinical practice</td><td>Changes practice to minimise error in individual practice</td></tr>
</table>

3.2 Task Analysis and Task Support

<table>
<tr><td rowspan="2">1. Outline and explain the psychology of human error and error modes
2. Analyse the range of tasks in the clinical area and evaluate tasks types as skill-, rule- and knowledge-based
3. Apply knowledge of performance influencing factors and their effect on clinical error
4. Evaluate safety-critical tasks where support is required to minimise error and improve the quality of clinical practice</td><td colspan="2">Understands the diverse nature of tasks in clinical practice and how to translate this into quality and safety improvement</td></tr>
<tr><td>Categorise personal tasks systematically and identifies the potential for safety improvement</td><td>Ensures that safety-critical tasks are adequately supported in the specialty</td></tr>
</table>

3.3 Non-technical Skills and Clinical Practice

<table>
<tr><td rowspan="2">1. Use case studies to understand the effect of non-technical skills on clinical practice
2. Carry out evaluation of personal non-technical skills (communication, situational awareness, stress management teamwork and leadership)
3. Outline and explain the hierarchy gradient and its effects
4. Apply strategies to improve non-technical skills in the specialty</td><td colspan="2">Recognises and works to improve non-technical skills as a way to build safe clinical systems</td></tr>
<tr><td>Is aware of non-technical personal non-technical skills and their effect</td><td>Actively evaluates and works to improve non-technical skills in the specialty</td></tr>
</table>

3.4 Process Reliability and Safety

<table>
<tr><td rowspan="2">1. Explain the relationship between clinical outcomes and process reliability
2. Identify and map safety-critical processes against clinical goals
3. Create and apply metrics to assess process reliability and clinical outcomes
4. Evaluate and develop communication and feedback to clinicians to improve process reliability</td><td colspan="2">Uses knowledge of clinical systems and processes reliability to improve patient safety and clinical outcomes</td></tr>
<tr><td>Identifies processes that affect clinical outcomes</td><td>Measures and support improvement of safety- and quality-critical process</td></tr>
</table>

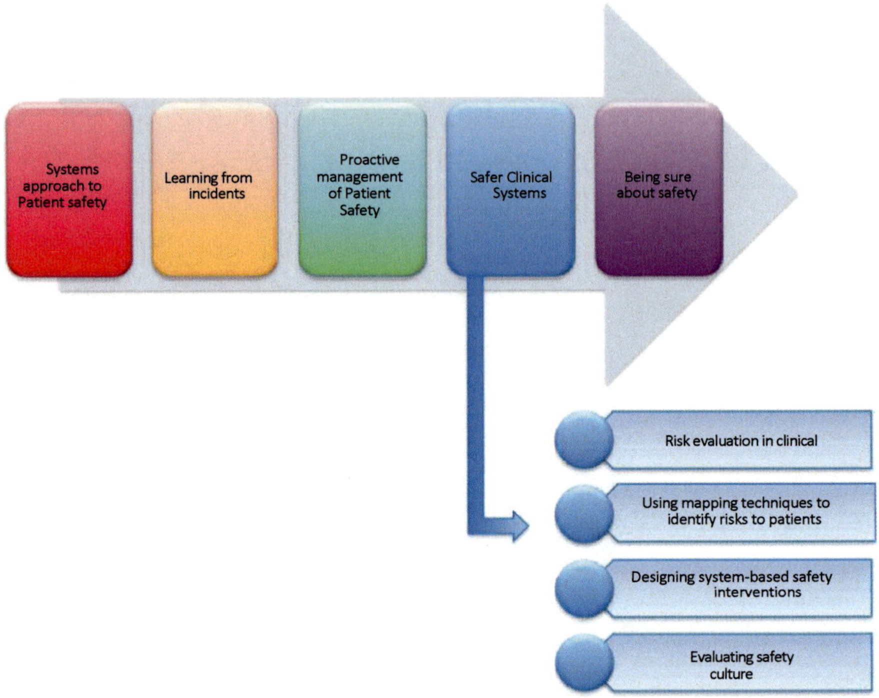

Fig. 10.5 Four key elements in Domain 4

10.10 Domain 4—Creating Safe Systems-(See 10.5)

Outcomes

Applies Proactive Risk Management in the Specialty to Create Safe Working Systems

Uses proactive safety techniques to prevent harm to patients; understands the strengths and weaknesses of safety interventions and the effect of contextual factors on safety; and evaluates dimensions of safety culture.

4.1 **Risk Evaluation in Clinical Practice**

<table>
<tr><td rowspan="2">1. Adopt a consensus-based approach to identify risk, with multi-professional involvement
2. Has knowledge of hazards and risks and can use a standard methodology to assess the risk to patients
3. Apply formal risk analysis of the defined area, patient pathway or treatment using failure mode and effects analysis (FMEA)
4. Identify proximal and systematic causes of potential failures and develop strategies to address immediate risks</td><td colspan="2">Uses both explicit and tacit knowledge of the clinical team in identifying and evaluating risk</td></tr>
<tr><td>Contributes to formal risk analysis in the clinical area</td><td>Leads to identifying risks using FMEA</td></tr>
</table>

4.2 **Mapping Techniques to Identify Risks to Patients**

<table>
<tr><td rowspan="2">1. Understands and apply process mapping to understand systems and to identify high-level risks to patents
2. Apply hierarchical task analysis to decompose safety-critical tasks and identify specific task risks
3. Take outputs from mapping techniques to structure improvement programmers in safety and quality and to manage risk
4. Use hierarchy task analysis as a tool to design goal-orientated safe clinical systems</td><td colspan="2">Develops a deep and detailed understanding of tasks designed to manage risk and create safety</td></tr>
<tr><td>Identifies areas of risk through process mapping and task analysis</td><td>Designs and implements safe clinical systems through goal-oriented HTA</td></tr>
</table>

4.3 Designing System-Based Interventions

<table>
<tr><td rowspan="2">1. Lead to the consensus-based evaluation of why things go wrong for patients
2. Outline and explain checklist design and use safety checklists appropriately
3. Outline and explain week and strong interventions in building safety
4. Apply the hierarchy of control to design and implement effective barriers to patient harm</td><td colspan="2">Uses system-based approaches to create strong preventative measures against patient harm</td></tr>
<tr><td>Contributes to consensus work in risk evaluation and solution design</td><td>Leads on introducing and monitoring barriers to patient harm</td></tr>
</table>

4.4 Evaluation of Safe Culture

<table>
<tr><td rowspan="2">1. Explain the key dimensions of reporting culture, just culture, flexible culture and learning culture
2. Apply the Manchester Patient Safety Framework (MaPSaf) as discussion and evaluation tool
3. Identify and apply formal safety culture evaluation instruments to the specialty
4. Encourage and support staff involved in safety incidents and ensure open and transparent responses in the specialty</td><td colspan="2">Uses a professional understanding of organizational culture to evaluate and support the creation of safety culture</td></tr>
<tr><td>Contributes to the assessment of safety culture and supports openness and transparency</td><td>Leads a multi-professional approach to assessing and developing a safety culture</td></tr>
</table>

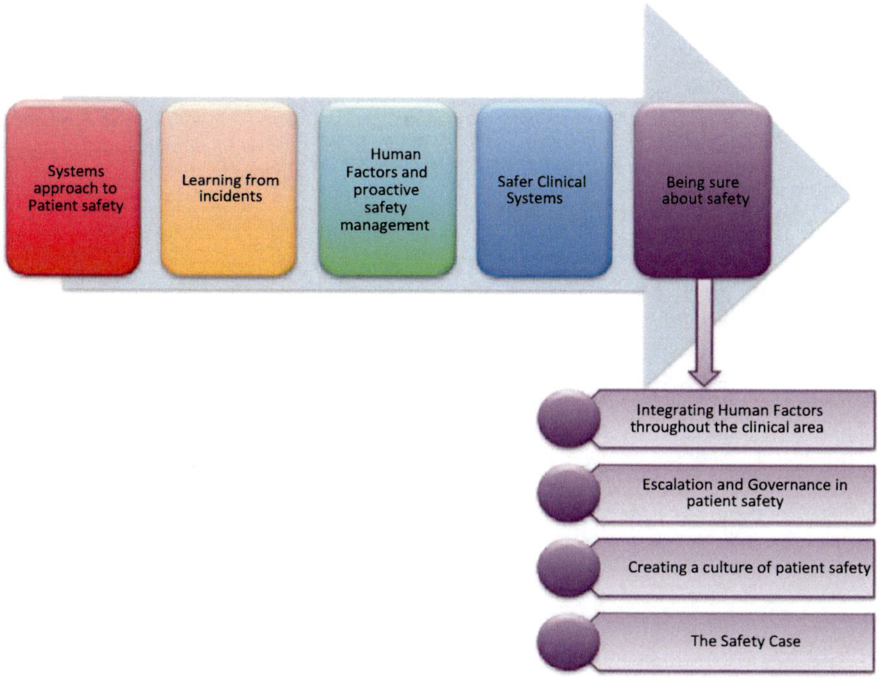

Fig. 10.6 Four key elements in Domain 5

10.11 Domain 5—Being Sure About Safety-(See Fig. 10.6)

Outcomes

Continually Monitors and Develops Patient Safety Through Human Factors and Systems Improvement

Uses proactive safety techniques to prevent harm to patients; understands the strengths and weaknesses of safety interventions and the effect of contextual factors on safety; and evaluates dimensions of safety culture.

5.1 **Integrating Human Factors**

<table>
<tr><td rowspan="2">1. Evaluate and ensure human factors integration through regular assessment against a formal system review checklist
2. Check safety-critical tasks and provide task support and usable, effective procedures for all clinicians
3. Identify, support and contribute to the design and implementation of safety-critical handovers and communications
4. Apply continuous monitoring of key risks and process reliabilities</td><td colspan="2">Ensures that human factors are a continuous focus of attention</td></tr>
<tr><td>Supports the use of human factors integration throughout the specialty</td><td>Actively identifies and develops human factors approaches to safety</td></tr>
</table>

5.2 **Risk Assessment Escalation and Governance in Patient Safety**

<table>
<tr><td rowspan="2">1. Understand and use specialty clinical governance meetings to review risks and identify residual risks
2. Justify and apply the risk management strategies of avoid, transfer, mitigate, contain or accept
3. Populate the specialty risk register with current and residual risks
4. Escalate uncontrolled risks to the next level of the risk hierarchy and monitor response</td><td colspan="2">Adopts a professional response to risk management at specialty or practice level</td></tr>
<tr><td>Support the use of risk management systems and raises risks to be addressed</td><td>Monitors residual risk and ensures appropriate escalation and governance of risk</td></tr>
</table>

5.3 Creating a Culture of Patient Safety

1. Foster an open, multi-professional approach to patient safety using both reactive and proactive methods 2. Develop or adopt techniques such as PRIMO, sharing lessons learned or use of huddles as cultural interventions 3. Use case studies from healthcare and other industries to ensure a continuing focus on safety management 4. Promote the principles of measuring and monitoring patient safety	*Places patient safety centrally in the specialty or practice*	
	Contributes to a safety culture through the use of case studies and safety interventions	*Contributes to a safety culture through the use of case studies and safety interventions*

5.4 The Safety Case

1. Create a safety case for the specialty or practice with a defined scope, evaluation of safety level, description of risks, risk control measures and residual risk 2. Apply the safety case as a tool to measure and monitor the safety 3. Use the safety case to address residual risk through improvement activities 4. Develop the use of safety case as a tool in governance and regulatory compliance	*Creates and applies a safety case*	
	Is aware of and supports formal safety management through a safety case	*Contributes to a wide understanding of safety by leading in applying a safety case*

References

Baddeley, R. (2017). An uncomfortably mismatch between control and responsibility. http://blogs.bmj.com/bmj/2017/11/robin-baddeley-the-hadiza-hawa-garba-case-highlights-an-uncomfortable-mismatch.

Chartered Institute of Ergonomics and Human Factors. (2018). *Human factors for health and social care*. London: Redactive Publishing.

Frankel, A., Haraden, C., Federico, F., & Lenoci-Edwards, J. (2017). *A framework for safe, reliable and effective care*. White paper. Cambridge, MA: Institute for Healthcare Improvement and Safe and Reliable Healthcare.

Hollnagel, E., Braithwaite, J., & Wears, R. L. (2013). *Resilient health care*. Surrey, England: Ashgate Publishing Ltd.

Kellogg, K. M., Hettinger, Z., Shah, M., Wears, R. L., Sellers, C. R., Squires, M., & Fairbanks, R. J. (2017). Our current approach to root cause analysis: Is it contributing to our failure to improve patient safety? *BMJ Quality & Safety, 26*, 381–387.

NHS England and Scotland. (2018). *Systems thinking for everyday work*. Edinburgh: NHS Education for Scotland.

Oliver, D. (2018). Are comparisons between acute healthcare and the aviation industry insidious? *BMJ, 361*, k2203 doi.

Patient Safety Hearing. (2018). *A patient-safe future*. London: Patient Safety Hearing.

Spurgeon, P., Flanagan, H., Cooke, M., Sujan, M., Cross, S., & Jarvis, R. (2017). Creating safer health systems: Lessons from other sectors and an account of an application in the Safer Clinical Systems Programme. *Health Services Management Research*, 1–9. (OCO).

Appendix A
Learning from Incidents

For illustrative purposes, we provide here examples of knowledge skills and behaviours relating to capabilities in Domain 2.

A.1 Learning from Incidents—Knowledge, Skills and Behaviours

In the context of creating safe clinical systems		
Knowledge	*Skills*	*Attitudes and behavior*

A.1.1 Investigating Patient Safety Incidents

Demonstrates knowledge of:	Demonstrates the ability to:	Demonstrate:
• Why patient safety incidents and near misses should be investigated • The influence of systems and human factors in creating the conditions for clinical errors	• Contribute to and conduct a professional standard of incident investigation • Use open or "cognitive interviews" to build an in-depth understanding of the events	• A willingness to embrace a multidisciplinary approach to investigating incidents with respect for all contributions • A focus on using investigation to achieve

(continued)

P. Spurgeon et al., *Building Safer Healthcare Systems*,
https://doi.org/10.1007/978-3-030-18244-1

(continued)

• Potential biases in attributing causes in investigations	• Write an investigation report which includes a timeline, an analysis of care delivery problems, and causal and contributory factors • Ensure that incident reports include robust recommendations for change	higher levels of safety for future patients

A.1.2 Designing Systems-Based Solutions

Demonstrates knowledge of:	Demonstrates the ability to:	Demonstrate:
• The range of possible safety interventions in a system and their relative effectiveness • The importance of system change or redesign in preventing future harm	• Use investigative team's and speciality clinical team's skills to develop realistic interventions for patient safety • Choose robust, systems-based interventions over weaker training or administrative interventions	• Openness and respect for contributions from all sources when developing interventions • Determination to effect change for patients as well as completing necessary records in investigations

A.1.3 Preventing Human Error

Demonstrates knowledge of:	Demonstrates the ability to:	Demonstrate:
• Human error as a widespread phenomenon, even with clinical leaders and in clinical situations • Human error modes and the influence of contextual factors on each type • Examples of human error in clinical practice and interventions aimed at error prevention	• Create or modify clinical systems so as to minimise the possibility of error • Identify contextual factors such as distractions, interruptions, workload and process ambiguity that may affect error	• An acceptance that error will occur but that systems can manage it • A commitment to bypass immediate blame for error and to prioritise ways to prevent further error

A.1.4 Avoiding Blame and Creating a Learning Culture

Demonstrates knowledge of:	Demonstrates the ability to:	Demonstrate:
• The effect of a culture of blame on the open disclosure of risk and safety issues • The underlying principles of separating individual culpability or capability from systems-enabled failures in patient care	• Apply the Incident Decision Tree to discrete human failures • Share learning from incident investigations widely in the clinical speciality and the wider organisation	• Freedom from bias in incident investigation • A commitment to fairness and transparency when contributing to or leading incident investigations

Appendix B
Underpinning knowledge and Expertise to Support Syllabus Domains

The illustration below shows the developing themes in human factors, systems expertise, risk expertise and safety culture as they are covered in each sequential Domain.

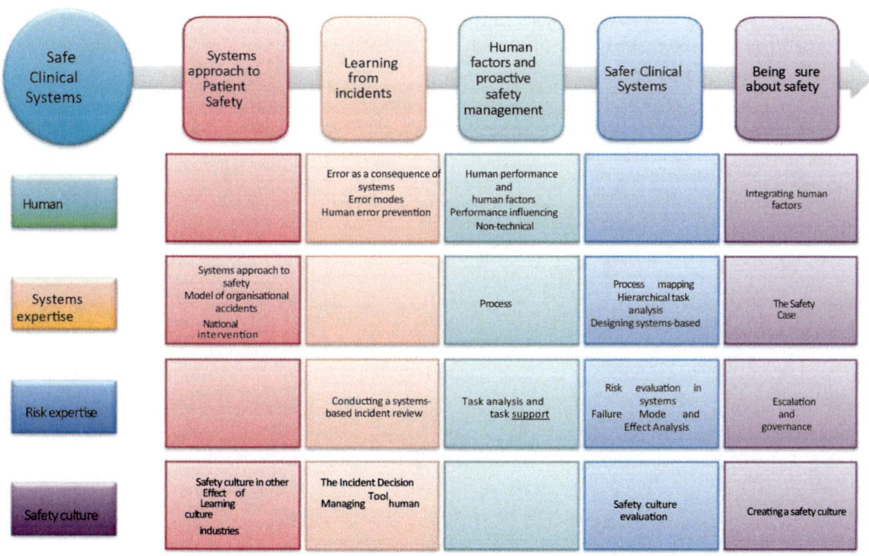

P. Spurgeon et al., *Building Safer Healthcare Systems*,
https://doi.org/10.1007/978-3-030-18244-1

Made in the USA
Las Vegas, NV
15 September 2026

7db8e365-e18e-4b14-9b64-3e59cde8027fR01